JOHN DEERE TRACTORS AND EQUIPMENT

VOLUME ONE
1837—1959

**Don Macmillan
and
Russell Jones**

American Society of Agricultural Engineers

2950 Niles Road, St. Joseph, Michigan 49085-9659 USA

About the
American Society of Agricultural Engineers

ASAE is a technical and professional organization of members committed to improving agriculture through engineering. Many of our 11,000 members in the United States and more than 100 other countries are engineering professionals actively involved in designing the farm equipment that continues to allow the world's farmers to feed increasing numbers of people. We're proud of the triumphs of agriculture and dedicated to preserving the progress for others. This book joins *The Agricultural Tractor 1855-1950, Farm Tractors 1950-1975,* and 17 other popular ASAE titles in recording the exciting developments in agricultural equipment history.

Acknowledgements

In a work of this size and complexity many people contribute to its success. Don Macmillan's association with Deere and Company and its products for 48 years led us to an important contributor to the book, the company's Archives and Leslie Stegh, Ph.D. The many rare photographs and illustrations provided have helped to widen the scope and interest of the book, and special thanks are due to Les and his team for making this possible.

An added input from the company was provided as a result of the sesquicentennial celebrations in 1987 at Waterloo, Iowa. Ralph Hughes, director of advertising, had taken many beautiful photographs and, in addition, the official photographs taken at that event were also made available by Don Huber. Both these gentlemen and others of their staff gave valuable contributions to the prime objective of a quality presentation, and for all this help our thanks are due.

The Two-Cylinder Club, and Jack Cherry in particular, were responsible for arranging the Waterloo show, and several photographs and much advice were received from him. Don Macmillan, having known his father, Lyle, since 1947, feels it is a great bonus that Jack has kept his records and continues his interest in the history of John Deere tractors.

Len Lindstrom, the project editor, has shown infinite patience in making the principal author's English understandable to the vast majority of readers. His knowledge of Deere products and natural editing expertise were most valuable, and a genuine rapport developed between us.

The pictorial nature of the book meant it was essential to achieve an attractive overall layout, and the addition of Dudley Gray, a talented free-lance artist and designer, to the production team has, we believe, resulted in a pleasing finish.

Contributions from Caterpillar Inc. and their Tom Biederbeck, with early Holt combine harvester photographs and information are acknowledged; to the University of California at Davis and the Hal Higgins section in their library for providing photographs to supplement those from Deere, to Bob Lovett for allowing extracts from his Export and Overseas Recollections, to Charles Wendel for the photograph of his beautiful Waterloo Gasoline Traction Engine Co. stationary engine and Jim Peters for the colorful Waterloo Boy poster, our grateful thanks.

Leading engineers from the membership of the American Society of Agricultural Engineers made helpful suggestions, and a vast number of friends, fellow collectors and others associated with this fascinating industry and hobby have contributed to this major work and made it possible. We would like to acknowledge all of them and express our grateful thanks.

Lastly to the whole ASAE team in St. Joseph a very special thank you—to its leader Donna Hull for her patience with authors seeking perfection, a state not easily attained; to Bill Thompson who, among all his other duties, spent endless hours with the same objective, and whose artistic talent contributed so much to the successful completion of the book; to Terry Burford, the maid of all work who handled all the word processing Stateside; and Dee Gunn who provided skill and control in the area of standards and flow editing. Truly this was a team effort by all concerned. Very well done, all of you, and many thanks to everyone.

Don Macmillan,
Russell Jones

Preface

Just as the majestic oak grows from a single acorn, so the John Deere "long green line" of farm equipment descends from a single walk-behind plow. The impact of this seemingly simple implement was far-reaching. It was central to the agricultural development of the Midwest, central to the development of Deere & Company, and central to the development of the western Illinois-eastern Iowa area where many of the John Deere factories are located today.

John Deere, the blacksmith, could not have foreseen the consequences of his imagination and skill at the anvil and forge. His interest in 1837 lay in resolving a literally sticky local problem involving the rich black prairie soil of western Illinois. Even a small amount of moisture made this soil adhere like putty to the moldboards, or earth-turning plates, of the cast iron and wooden plows then in common use. Seedbed preparation was impeded because farmers had to stop constantly to scrape accumulated soil from the moldboards.

John Deere figured out a solution to this problem. He forged a plow whose steel share, or cutting edge, was joined to a smooth wrought-iron moldboard. He formed the moldboard into a particular curve, or shape, so that even damp earth slid off. The plow self-scoured, or self-polished, in other words. Thus, pioneer farmers could plow faster. They could do more work. Productivity improved.

By 1842, John Deere was building 100 plows a year in Grand DeTour, Illinois, and couldn't keep up with demand. In 20 years, the John Deere Plow Factory, now in Moline, Illinois, was manufacturing nearly 15,000 plows yearly. Every year brought modifications and additions to the John Deere plow line: improved designs, new sizes—each specially designed to meet differing farmer needs or desires.

In this book the authors describe how the John Deere product line expanded. Some of this expansion was by invention, some by adaptation or modification and much by acquisition. This first volume covers the years between 1837 and 1959. During this span, farm equipment underwent more change than in the previous history of mankind. All over the world enterprising inventors and manufacturers were making tremendous improvements in the equipment farmers had used for thousands of years. During this period farmers got off their feet and rode on the equipment with which they farmed. The principal power source—at least in the U.S. and other industrialized countries—changed from animals to the internal combustion engine. Planting of seed changed from hand sowing to mechanical planters, listers and drills.

Harvesting of grain changed from hand scythes, grain binder and thresher to self-propelled combines. Hay making changed from hand forking into haymows to compressing hay into compact, wire-tied bales.

During the first half of this 122-year period, the John Deere brand was only one of many farm equipment brand names commonly found on farms. By 1959, however, the John Deere product line had become one of the most diverse and best known in the industry. This book, through photographs from the Deere & Company archives and the authors' private collection, illustrates how the John Deere product line expanded and diversified. Most of the changes parallel developments in other farm equipment manufacturers' product lines. Thus, one can envision chronologically how and when farm production was transformed mechanically, and why millions of farmers and their offspring were released for work off the farm.

Boyd C. Bartlett,
former president, Deere & Company

Contents

PART II—PRODUCT REVIEW

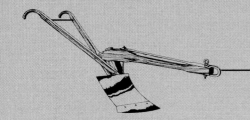

PART I

COMPANY HISTORY

1837–1887 the plow era

The growth from a blacksmith's shop to a major manufacturer of tillage equipment. In early years, plows were the most important product with cultivators and harrows representing a third of unit volume by 1883. The founder, John Deere, was recognized as an industry leader. He died in 1886.

John Deere the Founder

The man John Deere, whose name was to become known worldwide, started his business activities as a rural blacksmith. Deere & Company and the range of John Deere products are today identified as market leaders. Product quality, innovative products developed to meet user needs, and astute marketing and financial management have been the basis of this industry success.

It was in the year 1837 that John Deere first applied the yardstick of quality, "I will never put my name on a plow which does not have in it the best that is in me," which was to be the key to the company's growth. His business served pioneer farmers whose major problem was the breaking up and cultivation of the rich but sticky soil of the Mississippi Valley.

Important events may not seem so at the time. What happened in Grand Detour, Illinois when John visited the sawmill belonging to Leonard Andrus and noticed a large blade, made of Sheffield steel, lying broken and cast aside, was one such occasion. From its use in sawing wood the blade shone, and the idea occurred to blacksmith Deere that this steel might be just the material that would allow the sticky soil of the Mississippi Valley to scour a plow's moldboard. The plows in use at that time would not cut and clear themselves.

As a result of this chance occurrence John Deere built the first successful steel plow, with a smooth moldboard shaped to cut cleanly and overcome this soil packing problem. It was from this small beginning that today's multibillion-dollar "Long Green Line" company came into being.

The Hinton Painting of John Deere and his first steel plow

John Deere was born on February 7, 1804, in Rutland, Vermont, to William Rinold Deere and Sarah Yates Deere, he an emigrant merchant tailor from England and she the daughter of a British soldier who had stayed on after fighting the Yankees in the Revolutionary War.

John was eight when he lost his father. He grew up in Middlebury, near Rutland, where his mother carried on her husband's tailoring business. Starting work with a local tanner, John attended Middlebury College briefly, but was always a practical rather than a theoretical man. He soon decided to apprentice himself to a Captain Benjamin Lawrence to learn the blacksmith trade.

In the early half of the nineteenth century the blacksmith was manufacturer as well as mechanic, creating many of life's necessities. Everything made of iron was within his scope, and young Deere was fortunate to have in Capt. Lawrence a strict master, but the kind for a boy to have. Skilled workmanship was his master's creed, and soon it became his own delight.

Completing his apprenticeship in 1825, he hired himself out as a journeyman to two blacksmiths in Middlebury, and in 1827 married Demarius Lamb. His mother had died the previous year.

He earned quite a reputation for himself making hay forks, polishing the tines "until they slipped in and out of the hay like needles," and his shovels and hoes were "like no others that could be bought... scoured themselves of the soil by reason of their smooth, satiny surface"... an augury for the future indeed.

During the next several years he, his wife, and their steadily growing family moved several times in Vermont, end-

The reconstruction of John Deere's blacksmith shop in Grand Detour, Illinois

A replica of the original 1837 steel plow, part of Deere & Co.'s historical collection

ing up in Hancock in 1833. One of the town's residents recalled that Deere had made a log chain from discarded scythes. While the chain was not a thing of beauty, "no two links being of a length or the same size... it had been a joy to its owners, for after more than fifty years' use it had never broken." This was another pointer to the future.

The Village Blacksmith

Times were hard in Vermont. So in November 1836, one year after another Vermonter, Leonard Andrus, had moved west to Illinois, John Deere decided to follow him to Grand Detour "by canal and the lakes to Chicago and thence by stage coach." The small village was on the Rock River about 100 miles west of Chicago, and here John Deere set to work as a blacksmith, there being no other within many miles.

Legend has it that he arrived with $73.73. Be that as it may, his wife, three daughters and two sons arrived to join him in early 1838.

3

One son was Charles Henry, who had been born in Hancock some three months after Deere's departure west.

One of his first jobs was to mend a broken pitman shaft in that same sawmill of Leonard Andrus, and thus was started a business association which lasted until 1848. It was in this sawmill that the broken saw blade – probably one of the familiar up-and-down saws rather than a circular blade as was popularly suggested – gave John Deere his idea, an ingenious one, to use the smooth steel of the saw blade, itself highly polished, to form the smooth surface for his plow's share and moldboard.

The Steel Plow

Different versions of what actually happened when the new plow was complete and taken across the river to Lewis Crandall's farm exist, and are told in the book *John Deere's Company,* but one thing is certain – the new steel plow was a great success and resulted in an order for two more to be made like it. It was truly "The Plow that Broke the Plains." From this small beginning the manufacture of plows increased until, by the time of the move to Moline some ten years later, 1000 plows a year were being produced.

Initially the steel for the plows came through Naylor & Co. in New York from Sheffield, England, as none was available in the United States. "The first slab of cast plow steel ever rolled in the U.S. was rolled by William Woods at the steel works of Jones & Quigg in Pittsburgh in 1846, and shipped to John Deere, of Moline, Illinois." While dates don't quite match, since Deere was still in Grand Detour that year, it is rec-

The Smithsonian's original 1838 John Deere steel plow

ognized that it was in fact into the 1850s before the steel from Pittsburgh matched in quality that from Sheffield. The smoothness demanded by John Deere was one quality that set his plows apart from those of his competitors.

After informal business arrangements between the two, Leonard Andrus and John Deere went into formal partnership on March 20, 1843, and agreed that their resulting two-story plow works should operate under the name "L. ANDRUS." This agreement was to last for three years. Production of plows gradually

The New John Deere Plow Factory in Moline, Illinois, in 1859

increased from 75 in 1841 to 100 in 1842 and to 400 in 1843. As a result the works had to be extended, and steam power was added to run the machinery.

Developments at Moline

By 1848 John Deere noted that the new railroad being planned would bypass Grand Detour, and with production of plows rising continually, the question of transport was becoming critical. He decided therefore to dissolve his partnership with Andrus, and move to Moline, Illinois. He took with him Robert N. Tate, who had been employed to install a new steam engine in the Grand Detour works, and they formed a new partnership of Deere & Tate. The new firm began manufacturing plows in Moline, which though still only a village was larger than Grand Detour, and was located on the Mississippi and Rock Rivers with extensive transportation facilities.

Tate, an Englishman, moved to Moline as soon as the new agreement was signed on June 19, 1848. Wasting no time, he began constructing a new 24 × 60-foot blacksmith shop, and within a few weeks the building was finished. On August 31, the machinery in the building "moved for the first time," and was in production by mid-September: "September 26 – finished the first ten plows!"

By introducing John Gould into the partnership in late 1848, John Deere shifted his own role to arranging sales and transportation of the finished product. Making the plows was now supervised by Tate, and keeping the accounts was left to Gould. Business picked up in 1849, and in the first five months of that year no less than 1200 plows were produced. As a result the works was extended with a 2-story 30 × 80-foot shop, built massively to take heavy machinery.

Gradually the types of plows manufactured became more and more diverse. By the spring of 1851 stubble plows were available in 9″, 10″, 11″, 12″, and 14″ widths, and breakers in 16″, 18″, 21″, and 23″. River transport was a major benefit; for example, 439 plows were shipped down the Mississippi and up the Missouri, many destined for the Indian communities.

In a move ahead of its time, the partnership started to market the Seymour grain drill, a 6′ wide 9-row drill selling for $80. Its great advantage was that the driver could see if any of the tubes were blocked. It had won premiums at various fairs, but only one was sold in 1851, although seven were in stock. Advertising of this drill, although stopped after 1853, indicated a readiness to diversify further when the time was right.

Rapid Expansion

By 1851 production of all types of plows had reached 75 a week. The following year James Chapman, an attorney who had married Deere's daughter Jeanette, joined the group. Soon afterward, the partnership with Tate and Gould was dissolved and the new firm from 1853-1857 was called simply "John Deere." At the same time Deere's son Charles, having completed his education, joined his father. These five years proved to be prosperous ones. From approximately 4000 plows in 1853, production rose to over 13,000 by 1856.

Company production for 1857 according to the publication *The Cultivator* was 13,400 in all, as follows:

800 Large Breakers   18″ – 24″ furrow width
1300 Small Breakers   12″ – 16″ furrow width
9000 Stubble Plows   12″ – 14″ furrow width
1000 Corn Plows
300 Michigan Double Plows
100 Double- and Single-Shovel Plows and Cultivators
900 Other Items

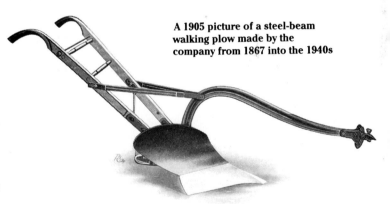

A 1905 picture of a steel-beam walking plow made by the company from 1867 into the 1940s

Up to 1857 Deere's plows were known as "Center-Draft," but after that date they were called "Clipper" and so advertised.

Yet another new product, ox yokes, appeared in the company's advertising in 1857… "fine specimens, and so far as we are able to judge, correctly made. This is an important matter – the making of Ox Yokes. There are some who are professionals that know nothing of the art of making a yoke in which oxen will work with ease."

The late 1850s saw another development which was to have far-reaching effects on farming in that era. Railways had begun to extend all over the industrial countries of the world, and it was only natural that somebody should begin to consider the possibilities of steam power for plowing and cultivation. The British answer had been to use one or two engines with cables and winches to haul plows and cultivators across the fields, and a few of these monsters had been exported to the United States. With the large fields available, this idea did not prove popular. American farmers wanted a self-propelled steam engine to pull the implement directly, as their mules and horses did.

One Joseph W. Fawkes from Pennsylvania captured the nation's imagination at the state fair in Centralia, Illinois, in 1858, by plowing with his 30-hp vertical-boilered engine "Lancaster." He made contact with John Deere, and it would appear that his engine was taken to the Moline works, where eight John Deere plows were combined for him to make further attempts the following year. After showing at the Illinois State Fair, the outfit was taken to the U.S. Agricultural Society's contest in Chicago, where it gained the championship and was awarded the Grand Gold Medal of Honor "for that machine which shall supersede the plow as now used... with the greatest economy of labor, power, time, and money."

John Deere Sells Out

In 1858 John Deere sold out his interests in the company to Christopher Webber, his son-in-law, leaving him and Charles Deere owning the company. Already the 21-year-old Charles was looking to the future. Notes in his private memorandum book refer to "see the corn sheller made in Chicago, it is an apple grinder too... Whittemore & Belcher & Co., Chicopee Falls, Mass. make a good corn sheller, the No. 10 is best size... see Morrison, Pittsburgh about mold castings."

But there were difficult years to weather before he could put into practice his vision of the future.

In May 1860, Charles' sister Emma was married to Stephen H. Velie, a name that will appear over the years as this story unfolds. Two years later Charles married Mary Little Dickinson from Chicago. Velie joined Charles in the business in 1863, when it was a small regional operation. He and Charles worked closely together, and when he died in 1895 the company had risen to a prominent place in national esteem.

Deere & Company Reconstituted

Although John Deere had sold his interest in the firm to Webber in 1858, late in the Civil War the company was again reconstituted as Deere & Company, with John and Charles sharing equally in the partnership. In truth John had never been far away from his beloved plow works. Early in January 1864, John Deere obtained his first patent, relating to molds for casting steel plows, and followed it with a second a few months later. His third, in 1865, was a complicated one relating to the interrelationship of share, moldboard, and landside, and the method of securing same.

In 1863, however, with no mention anymore of ox yokes in the company's catalog, a cultivator, the Hawkeye, was added to the line. This implement was the first from Deere adapted for riding, and thus a forerunner of the famous Gilpin sulky plow introduced in the seventies. The Hawkeye won the first premium at the Iowa State Fair in its first year and rapidly achieved acclaim. Some 500 were sold in 1864 because of the ease of operation, all the shovels being raised clear of the ground by pressing one's feet on

An artist's impression of an early gang plow with steam traction engine. Deere's first gang plows were built in 1858-1859

John Deere's first departure from plow manufacture – the Hawkeye sulky corn cultivator introduced in 1863

a pair of treadles, thus leaving the hands free to guide the team.

John Deere Incorporated

The many partnerships in which John Deere had participated finally came to an end in 1868. On August 15, Deere & Co. was finally incorporated with John, Charles, Stephen Velie, and George Vinton as shareholders. They were joined a year later by Charles V. Nason and Gilpin Moore. In its first full year of trading, the company sold 41,133 plows, harrows and cultivators, with a turnover of $646,563. In the following year a No. 1 Clipper plow cost the farmer $26, a 12″ Breaker $35, and a 24″ Breaker $51. In this year, 1869, a new system of marketing their products was introduced by the company at the instigation of Charles Deere. A separate and distinct "branch house" was organized in Kansas City with Charles Deere and Alvah Mansur as the two responsible, and the new company was called Deere, Mansur & Co. Over the next 20 years four other branches followed in St. Louis, Minneapolis, Council Bluffs/Omaha, and San Francisco.

During the mid-sixties Deere had started to market the Black's gang plow, the first walking gang, but in 1874 they came on the market with their own Deere gang, available with double-and triple-bottom models. The machine which became so well known

This painting shows John Deere watching a single-furrow Gilpin sulky plow at work

7

that it fundamentally influenced the growth of the company was the Gilpin sulky plow. It was introduced in 1875 and immediately patented.

An Osage, Iowa implement dealer wrote to Charles Deere in 1876 eulogizing the Gilpin's ease of handling. "I took it to our county fair and put it into a trial with the Crosley. We selected a piece of ground, one end of which was good smooth prairie, while the other end was very rough, and a low piece of breaking ground that had gone back and was thickly covered with weeds. To satisfy the many spectators present that the plow was easy to handle, I got a lady who had never run a plow before to mount the Gilpin, while a man of a good deal of experience ran the Crosley. The Gilpin was run at depths from 4 to 11″ deep, and in every instance it done its work splendidly. Much better than any man could do with a hand plow. The difference in the work done by the two plows was so great that anyone could tell the difference nearly as far off as they could see the field. The Gilpin buried the tall weeds entirely out of sight, while the Crosley left them sticking out of the ground, and did not do as good work anywhere."

International Recognition

The Gilpin won so many firsts in competition all over the Midwest that the company advertising proclaimed it was "victorious in all points." Charles Deere decided, in view of the plow's success, to enter it in a much grander contest, the famous Paris Exhibition of 1878 at Petit-Bourg, France. The greatest interest was shown in the Gilpin; it won the contest easily and was presented with one of seven special prizes, a Sevres vase, valued at 1,000 francs. The Paris

An attractive advertisement for a New Deal gang plow with three wheels first introduced in 1884

A 3-wheel New Deere gang plow at work in July 1905

Exhibition was a major boost not only for Charles Deere's morale but also to the reputation of two of the most important products of the company, the Gilpin sulky and the Deere gang.

In 1881 the Gilpin was improved in a major way by the development of a "power-lift," allowing the plow bottom to come out of the ground by simple pressure of the hand on a catch which connected to the land wheel. Few people would argue with the company's claim that it was "The King of the Riding Plows." During his 23 years with Deere, Moore contributed a great deal, including 31 patents processed in his name.

In 1884 the company adapted the 3-wheel configuration for its gang plows, all the previous models having only two wheels, and the following year made a further improvement with the addition of a depth lever. These 3-wheel plows were called New Deal Gangs. By 1889 New Deal plows were available as single furrow walking or riding, and from 2- to 6-furrow gangs with 6″, 10″, 12″, and 14″ furrow widths. Since the 2-furrow plows needed from four to six horses to pull them, other forms of power were needed for the larger units. The increasingly popular steam traction engine met this need for greater power.

Unnoticed by the company but destined to have a considerable effect on its later expansion, the combine harvester was becoming a practical proposition at this time, in distant California where Benjamin Holt had put his famous Link Belt-driven combine into production in 1886. But more of that later... the 1870s had seen other developments in the company's line.

In 1876 the Hawkeye cultivator had been replaced by the Peerless and the Deere Riding cultivators. A

Scotch harrow with 40 teeth had also been added. The following year Deere & Mansur, a separate company in Moline, had introduced a corn planter, and in 1879 they came out with a corn stalk cutter. These machines, although built by a separate concern, were marketed by the various Deere branch houses, and this idea was the forerunner of marketing principles adopted later. The company was ever mindful of the needs of its customers, and introduced new products accordingly.

50 Years of Rapid Growth

A look at the machines marketed in the early eighties illustrates the company's growth as it approached its first fifty years in business.

Year	Walking Plows	Gilpin Sulkies	Spring Cultivators	Shovel Plows	Harrows
1879	40,544	5,497	4,862	5,328	6,727
1880	39,464	5,964	10,067	5,659	5,938
1881	45,171	5,773	15,464	9,018	6,326
1882	50,025	7,704	16,777	9,645	7,393
1883	48,858	7,841	13,818	9,198	14,604

In 1881 Deere & Mansur brought out their new "Deere" and "Moline" planters for use with any standard check-rowers. They both utilized the Deere Rotary Drop system, and despite patent arguments continued in production. A company advertisement of 1883 noted "The Moline worked very effectively – each of the six chambers in the rotary disc would always drop a kernel when the opening appeared, and there was apparently very little cutting of the corn kernel, which, if it happened, destroyed the kernel and caused a skip in the planting. They undeniably increase the crop and prevent insects from disturbing the corn when small."

Another "small jump forward," rather than a positive leap, occurred in 1885 with the introduction of a sulky rake, a small venture into the harvesting world, although company catalogs still offered five competitor models as well.

John Deere Dies

On May 17, 1886, the founder of the company died at the age of 82.

As concluded by Wayne Broehl, Jr. in *John Deere's Company,* "In retrospect, John Deere's contribution to his company and to the agricultural industry all over the world emerges clearly. He was energetic and purposeful, moving forward despite ownership squabbles, business ups and downs, marketing battles, and countless other problems. He was not himself a major inventor; on the other hand there is ample documentation of his highly productive role as an adapter and marketing innovator. He was not a financier; indeed, his lack of knowledge and judgement about money matters almost lost him the Company more than once. Nor was he always the diplomatic leader; often his temper exploded over small slips by others. But he did have a knack for organization, an abiding concern for quality, and a feeling for the role of the agricultural equipment industry in America's growth, that made him a preeminent producer and distributor of farm machinery. Above all, he was a charismatic leader, a man of great bearing, commanding integrity and loyalty, and he must have instilled more than a modicum of leadership qualities in his successor-son, Charles Deere."

Portrait of John Deere in later life

9

1887–1912 expanding the line

A period of expansion and the development of a full line. In the years covered, the company moves from a tillage specialist into the manufacture of corn drills, planters, wagons, binders, farmyard equipment, and similar farm machinery. Rapid expansion in volume across the range is achieved from a growing number of manufacturing plants. Leadership is in the hands of Charles Deere and, later, William Butterworth.

The Line Is Extended

During the first 50 years of the company's development, it had concentrated on tillage equipment, but it was the next 25 years that were to see Deere & Co. offer a full line.

The blacksmith's forge had become the John Deere Plow Works, the company's only factory. By 1870 the buildings covered 90,000 square feet. In 1882 this had risen to 9 acres, and by 1904 to 15 acres with a further 35 acres of floor space. Electric lighting had been in use since 1882. In 1880, the Works produced 180 types of plow; by 1903 it was the nation's largest producer of cultivators, making one-fifth of all those sold.

During the 1880s Deere had developed links with other companies that were producing the farmer's favorite means of transport, buggies for his journeys to town and wagons for farm hauling. Deere had acquired a one-third interest in Deere, Wells and Company, which had started in 1853 in Moline making wagons. Through the connection with Mansur, Deere had links with Mansur & Tebbetts Carriage Co., which was assembling and marketing buggies.

In order to satisfy the "great bicycle craze," the Minneapolis branch launched the sale of the new "lowdown" or safety bicycle, and by 1895 was selling the Deere Leader, Deere Roadster and Moline Special models. In August of that year they held the "Deere Road Race" over a 20-mile course, with over 100 entries. The peak period for bicycle sales was from 1892 until 1898; by the turn of the century the company was virtually out of the bicycle business.

An interesting picture of the Plow Works in Moline

Charles Deere plowing with two horses and a steel-beam walking plow

Studio portrait of Charles Deere

Farm Machinery Boom

The end of the nineteenth century saw a boom in the farm machinery business, which continued into the first decade of the twentieth. An indication of this is best illustrated by the number of tillage implements produced by the company, shown in the two tables below:

Product	1899	1909
Small Cultivators	207,171	469,696
Wheeled Cultivators	295,799	435,429
Disk Harrows	97,261	193,000
Other Harrows	380,259	507,820
Disk Plows	17,345	22,132
Shovel Plows	103,320	245,737
Steam Plows	207	2,355
Wheel (Sulky) Plows	135,102	134,936
Walking Plows	819,022	1,116,000

Growth in Production at the Moline Plow Works:
    1847 –   700 Implements
    1850 –   1,600
    1857 –   10,000
    1874 –   60,000
    1884 – 120,000
    1904 – 200,000

11

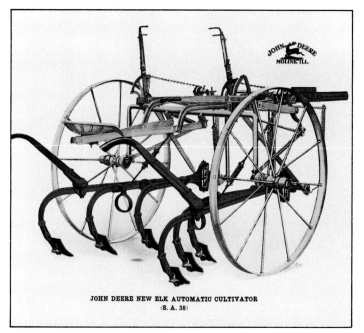

JOHN DEERE NEW ELK AUTOMATIC CULTIVATOR
(S. A. 38)

DEERE MODEL B DISC HARROW
Flexible Spring Pressure

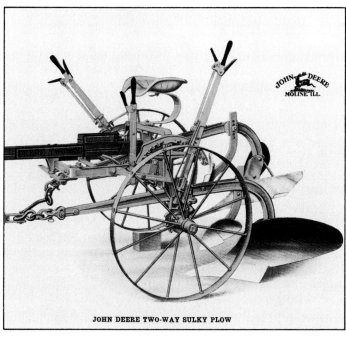

JOHN DEERE TWO-WAY SULKY PLOW

NEW DEERE LIGHT DRAFT SULKY

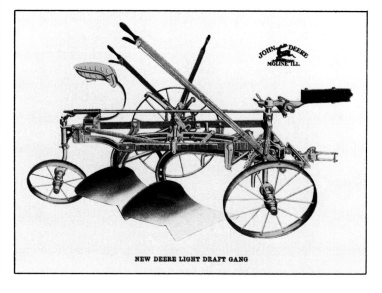

NEW DEERE LIGHT DRAFT GANG

Illustrations from the 1910 John Deere St. Louis Branch House Catalog

New Elk automatic cultivator
Two-way sulky plow
Model B disk harrow with flexible spring pressure
New Deere light draft sulky plow
New Deere light draft gang plow

Moves Toward a Full Line

Although the company was strong on its tillage side, it had still not expanded into the haymaking area in any significant way, a sulky rake being the solitary example in this field, nor into harvesting equipment at all. Having resisted a bid from a British syndicate, it was International Harvester which loomed large at the beginning of the new century, with their coverage of a full range of farm implements. It became increasingly clear that Deere & Co. should have a full line also, as well as a nation-wide distribution system.

Implement dealers and wholesalers had seen, at an early date, the advantages of listing a complete line of equipment. In fact, some of Deere's own branch houses had for 20 years been handling additional and non-competing lines such as Moline wagons. In 1900 the company started with new purpose on a quiet and careful selection of leading implements of all types then in general use.

As the investigation proceeded, they learned that Deere & Mansur (in which Deere & Co. then had only a minority interest) was making corn planters with few, if any, equals. They learned that the Syracuse Chilled Plow Co. had developed plows as well adapted to soil in the East as John Deere plows were suited to the heavy soil in the Midwest; that the Dain Manufacturing Co. (which since 1895 had sold much of its production through John Deere dealers) had established and was maintaining the superiority of Dain hay tools; and that wagons made by the Moline Wagon Co. were high in farmers' estimation because of the honest craftsmanship obvious from tongue to endgate.

They learned that the Kemp & Burpee Manufacturing Co., pioneer in the building of practical manure spreaders, was setting the pace in design and production; that the Van Brunt Manufacturing Co. was a leader in the grain drill field as regards both quality and volume; and that the Marseilles Manufacturing Co. had pioneered force-feed automatic corn shellers and had met such success that it added a second labor-saving tool, a grain elevator, which also met immediate acceptance.

The Search Completed

Ten years the search went on, and then came the great reorganization based on knowledge thus obtained. Late in 1910 the decision was made as to which farm implement plants merited membership in the proposed John Deere family of factories. Early the following year the merger took place, bringing together

Nichols and Shepard steam engine with 8-furrow gang plow in November 1910

various manufacturing companies as well as 22 selling organizations, to give Deere & Co. a full line of implements as well as a nationwide distribution system.

The new Deere & Co. was in no sense a combine or trust. None of the subsidiary manufacturers had been in competition with each other. The selling organizations were not to handle competing lines, and were not to have overlapping territories. The merger was intended to benefit not only the subsidiary organizations but also customers, who could now obtain practically any desired implement, as well as service and replacement parts, from a single dealer in any community where John Deere equipment was sold.

A great organization, sound, strong and respected, grew on the foundation thus laid as the twentieth century began. Head and shoulders above the other pioneer inventors whose ingenuity and purpose formed the member companies was John Deere, of course, but also on the honor roll are the names of Wiard… Moore… Kemp… Van Brunt… Adams… Dain… and many others.

The First Addition

The first acquisition of another company took place in May 1907, just before the death of Charles Deere, when the Fort Smith Wagon Co. was purchased. This company had been founded as the South Bend Wagon Co. and for some years used that name. South Bend founded the Fort Smith Wagon Co. in 1903 and business began in January 1904.

The plant was exceptionally well laid out, and the product from it continued to be known as the South Bend wagon until 1905. In the spring of that year the firm name was changed to Fort Smith, and so continued well after Deere & Co. had purchased the company. It had the distinction of having become a part of the organization before the period of general expansion.

It had been connected with Deere & Co. from its founding, the Dallas branch house patronizing it heavily from the start. Shortly afterward both the Omaha and Kansas City houses took up the line.

In 1907 the John Deere Plow Co. Ltd. was formed in Canada, Deere's main export market. Following the board decision on January 6, 1910, to purchase such other companies as were necessary to give them a full line, these other companies quickly followed into the Deere fold. An indication that all this would take place had occurred much earlier, when in 1899 the company became controlling owner of its previously independent branch houses.

A Model A horse-drawn manure spreader with extension sides – the spreader with the beater on the axle

Corn Shellers, Elevators, and Manure Spreaders

In 1908 the Marseilles Manufacturing Co. of Marseilles, Illinois, took a contract to supply the Deere & Mansur line of corn shellers. In this way a connection with Deere & Co. sprang up, which soon became the basis for the purchase of the company in 1910. It was renamed the Marseilles Co. and moved to East Moline.

A No. 1 hand corn sheller,

The original business was the outgrowth of the foresight of Augustus Adams, who was born in New York in 1806. He moved west and settled in Marseilles to engage in foundry work. This business soon grew to take in a line of elevators, and as time went on became by far the most famous corn sheller plant in existence.

To the output of this plant were added the spreaders of Kemp & Burpee of Syracuse, New York. This last plant had been the result of a successful spreader development by J.S. Kemp at Magog, Quebec, in 1877.

In 1880 Kemp & Burpee was incorporated in Syracuse. The business progressed very well, and their spreaders became justly famous. In 1902 they made a sales agreement with the John Deere selling organization. The famous "Success" spreader was produced at this time. As a result of this connection the plant became very closely allied to Deere & Co., and on December 1, 1910, it was taken over as an integral part of the greater company. It ran thus for only a short time, its lines being allocated to the other factories, and from this the Marseilles Company acquired the spreader business.

Haymaking Equipment

In 1881, Joseph Dain was in the retail furniture business at Meadville, Missouri. An idea came to him of reducing the cost of haying by the use of sweep rakes and stackers. His idea was opposed at first, as it was not thought that hay from the swath could be safely stacked. Dain, however, persisted. He sold his furniture business, and after securing patents in 1882 started a little shop in Springfield, Mo.

Changes of plant followed rapidly. In 1883 a plant was built at Armourdale, Kansas. The popularity of Dain's tools grew quickly, and in 1887 the Armourdale

Dain hay loader in a July 1920 hayfield

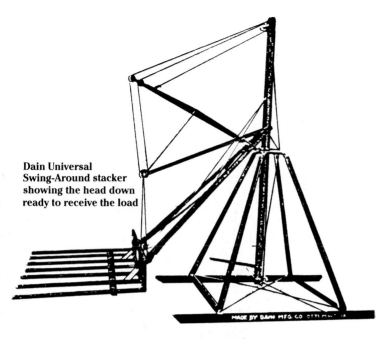

Dain Universal Swing-Around stacker showing the head down ready to receive the load

Corn Planters

Deere & Mansur was organized by Charles Deere and Alvah Mansur, who was at the time manager of Mansur & Tebbetts of St. Louis and Kansas City. The start was in 1877 with a capital of $25,000, and the purpose was to make corn planters and to fill out the line then made by Deere & Co. The corn planter business had up to this time been in the hands of George W. Brown, the pioneer in the line. In the fall of 1877 a building belonging to Union Malleable was taken over and work started for the next year. In October 1879 a building known as "The Quilt Factory" was taken over. The operation of a separate planter shop was unique with the company, for Deere & Mansur was, of course, very closely allied to Deere & Co., and the practice remained peculiar to them. This was due in large measure to the fact that the D & M corn planter always far surpassed others and always was The Corn Planter. Mansur died in 1898, but the company retained its separate identity until the reorganization of 1910-1911.

Chilled Plows

Syracuse chilled plow with slatted bottom

plant was outgrown and a new one built at Carrollton, Mo. The business was incorporated with a capital of $40,000 in 1890, with Dain as president. The first connection with the Deere interests was made in an alliance with the Kansas City branch house in 1895, and with Deere & Webber in 1899.

In 1900 the Carrollton plant also proved to be too small, so the company moved to Ottumwa, Iowa, where it has been ever since. Connections with the John Deere branch houses were growing steadily all this time, until in the great reorganization the Dain plant was felt to be essential to the business.

There was also an offshoot of the Dain interests in the Dain Manufacturing Co. of Welland, Ontario, which had been built in 1909–1910 to take over the Dain work for Canada. The Dain lines had been previously handled by an outside concern at Guelph, Ont. The company had 230 acres of land, which was of great use as the Canadian business grew.

On October 31, 1910, the Dain companies were formally taken over by Deere & Co. and Dain became one of its vice presidents, a position he retained until his sudden death in 1917.

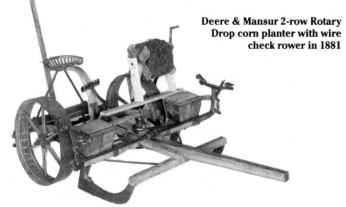

Deere & Mansur 2-row Rotary Drop corn planter with wire check rower in 1881

The roots of the Syracuse Chilled Plow Co. extended to the oldest plow building traditions of the country. In 1802, two years before John Deere was born, Thomas Wiard was building a wooden plow with an iron share at a shop in East Avon, New York. This business went on successfully, and his son William moved the plant to Ancaster, Ontario, about 1840, being then engaged in the making of an all-iron plow. In 1860 Henry, the grandson of Thomas Wiard, started the manufacture of plows at Oakfield, N.Y. He later developed the process of hardening iron which was used in the manufacture of chilled plows.

The Syracuse Chilled Plow Co. was started as the Robinson Chilled Plow Co. of Geddes in 1876. The following year sales were 8,457 plows, a large showing for a new company. In 1878 the business was rein-

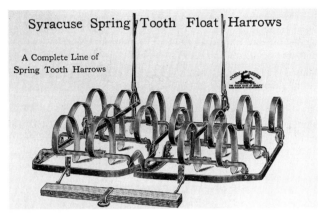

Syracuse spring-tooth float harrow

corporated under its new name. Henry Wiard designed the plow and was superintendent of the company for many years, but it fell to his son, W.W. Wiard, to become a vice president of Deere & Co., and thus a fourth-generation plowmaker.

Deere & Co., needing a chilled plow to fill out its line, found that the Syracuse plows were by far the best in the business. Accordingly, arrangements were made, and the Syracuse company was taken over in May 1911.

Just prior to this, in April, the Union Malleable Iron Co. had been acquired. It had been founded in 1872 near the site of the Deere & Mansur plant. It had a varied existence until Charles Deere took a controlling interest in 1894. The plant was moved to East Moline in 1901. With the greater merger it was thought to be of value to have the malleable parts supplied by a section of Deere & Co., resulting in the Union plant's takeover.

A Buggy Line

Taken from a 1900 buggy and surrey advertisement in the
Implement Trade Journal

The Reliance Buggy Co. factory was the outgrowth of the old Mansur & Tebbetts Implement Co. It started in the painting and finishing of purchased vehicles. From 1891 to 1902 it occupied rented quarters, and manufactured vehicles in them.

Deere & Co. acquired the Mansur & Tebbetts properties in 1899, and by 1902 the buggy business had increased to the point of building a new plant in St. Louis. With the organization of the greater Deere & Co., the plant was put on the same basis as all the other parts of the company.

Seed Drills

An early picture of the Van Brunt Manufacturing Company's drill works

The other major member of the new team and builder of seed drills was the Van Brunt Manufacturing Co. This company was founded at Mayville, Wisconsin, in 1860 by Daniel C. and George W. Van Brunt. Seven seeders were made in 1860, and then the company moved to Horicon in 1861. George dropped out and Daniel ran the business until his death at the age of 83 in 1901. Until 1890 they also built rakes and wagons.

The company made steady progress and 1890 saw the development of the first shoe drill, of which 3,000 were built. The single-disk drill of 1900 was produced in a 6,000 lot that year, and in 1912 the output was 28,000 machines.

Willard A. Van Brunt had been in the seeder business all his life, and having succeeded his father as president of the company, he became a director of Deere & Co. as well. He was directly responsible for the adjustable-gate forcefeed and the single-disk drill.

Because of the preeminent excellence of the Van Brunt line, Deere & Co. bought out the Van Brunt interests in June 1911, and was thus able to add to its line the best known seeders in the implement trade.

A 1911 Van Brunt combined grain and fertilizer disk drill—
the first of this type

Farm Wagons

Soon after John Deere moved from Grand Detour to Moline, one James First started a little wagon repair shop. While information as to the growth of this concern is sparse, nevertheless the company had developed sufficiently that in February 1872 it was able to incorporate with substantial capital. Maurice Rosenfield was the first president of the new company, which was therefore called Rosenfield & Co.,

but this name was later changed to Moline Wagon Co.

In 1883 it became associated with Deere & Co. in the formation of Deere, Wells & Co., subsequently to become the John Deere Plow Co. of Omaha. This was followed in 1886 by association in the foundation of the Kansas City and St. Louis branches. These transactions produced a close alliance with Deere interests, which continued steadily until, in January 1910, the wagon company entered the John Deere organization and became the John Deere Wagon Works.

Over the years the export side of the business had gradually increased. In 1903 export sales were $193,323 and by 1910 had risen to $561,042. As a result of this trade, the company decided to open their own offices in New York in 1908 in the name of the John Deere Export Company, an Illinois corporation. The site chosen was No. 17 Battery Place, and the company's East Coast forwarders in the 1980s are still at the same location. By 1911, however, it was decided to move back to Moline, and the new Export Department was set up there on April 1, 1912.

In 1906-1907, for a brief interlude, Charles Deere became president of the Deere-Clark Motor Car Co. and a few cars were sold, but this development did not persist.

Two horses in charge of two John Deere wagons

The John Deere Binder

One final piece of the jigsaw remains to be put in place to complete the picture in 1912. The need for a binder to round out the company's line was very apparent. The Acme plant was considered in 1908. In the meantime Canadian conditions became more acute. In 1909 the Frost & Wood plant at Smith Falls, Ontario, was discussed, but finding that this company had concluded an arrangement with Cockshutt, the Deere board directed in October 1909 that ways and means for manufacturing harvesters in Canada be reported on, and that work be immediately started on developing a harvester.

An early John Deere binder and four mules at Waco, Texas, in 1913. Photograph taken on the Westbrook Plantation with the foreman on the binder seat.

In the winter of 1909-1910 the John Deere binder was designed under the direction of Harry J. Podlesak, formerly with the McCormick works, and it was his excellent design which gave the Deere binder its ability to hold its own beside the older and more celebrated machines. In May 1910 A.C. Funk, superintendent of the Champion Works, came to Deere & Co. as head of the harvester department.

Seven or eight machines were tried out in the field in the summer of 1910, and during the winter and spring of 1911 five hundred binders were built in the old Root & Vandervoort plant in East Moline. The work up to this time had been carried out with the intention of ultimately making binders only in Canada.

In May 1911 W.L. Velie's resolution "That operations be started immediately in Moline instead of in Canada" gave the first definite start toward a harvester works in East Moline. During the summer of 1911 almost all of the 500 harvesters were sold in Canada.

Joseph Dain at this time first interested himself in the harvester operation, due to Funk's failing health.

During the latter part of 1911, Dain undertook temporary management of the harvester plant, and during that winter and the spring of 1912 two thousand binders were made in the same rented quarters. Plans were meanwhile being developed for a plant to manufacture harvesters, mowers, and rakes exclusively, and the present Harvester Works site was purchased.

Eventually, on Dain's insistence, W.R. Morgan was appointed the permanent manager of the new works. He had grown up in the harvester business with O.M. Osborne & Co., starting at Minneapolis in 1881. When International Harvester Co. took over Osborne, he was sales manager at their Auburn headquarters. He continued with International Harvester as their assistant sales manager in Chicago until his transfer to Deere.

The first section of the new Harvester Works was built in the summer of 1912, and the rest, including the warehouse and foundry, in the summer of 1913. The whole plant was occupied for manufacturing purposes throughout the 1914 season, when some 12,000 harvesters, and other machines in proportion, were produced.

Combine Harvesters

Tractors feature only marginally during the 25 years under consideration in this chapter, and that through the Waterloo connection, which did not occur for another six years. The combine scene, however, was developing out in California with the growth of the Holt concern.

Moore and Hascall had produced the first real attempt at a combine in 1834 in Michigan, but these early machines were not suited to the Midwest conditions, so a later version was shipped to California via the Cape for the 1854 harvest. This early Hiram Moore machine was burned, so in 1858 Strong & Taylor built a replacement for the farmer-owner, John M. Horner. During the period from 1859 to 1869 he operated three machines in the San Joaquin Valley. In 1869 William Marvin of Stockton, Calif., commenced experiments and won prizes for his machines at the Stockton fair, and by 1876 he had built two 10′ cut machines which operated for five or six years.

The Centennial harvester, so called because it was made in 1876, the Centennial year, was built by Young & Hoult. A few were built in the first year, followed by 13 in 1877 and about this number each year until 1890. These machines found a ready sale at $2,000 each. They too were push-type, with 24 horses or mules used 12 abreast. They had two wheels, the left one with gear drive to the thresher.

In the decade from 1876 to 1886 no less than 21 companies experimented, with varying degrees of success, in building combines in the Pacific region. Among a few of the better known names were Dave Young, who had left the partnership with J.C. Hoult; Houser; Matteson & Williamson; Myers; Minges; Gaines; Shippee; and Best. Best had started experiments in 1884, having moved from Oregon to San Leandro, and built and sold his first successful machine the following year. Due to its success he enlarged his plant and later designed and developed the first steam driven combine in 1889. The world's first self-propelled combine was built by G.S. Berry in 1886, with a 22′ cut.

In 1880 Daniel Houser had built the first factory used exclusively for the manufacture of combines, and by 1882 there were 25 machines working in the Stockton area.

1883 saw the Shippee company build 90 units, having acquired Davis Bros. of Oregon and Dr.

William Parish's patents, as well as the Warrish "King." During the next 12 months they also took over the Houser, Benton, Powell, Gratton, and Minges machines, enabling them to set up "Shippee's Stockton Combined Harvester and Agricultural Works."

The Holt Manufacturing Company

Most of the previously mentioned companies were gradually absorbed by the most successful California company, that of Holt Bros. All the previous machines had been gear driven, which was not very satisfactory with horse or mule haulage, since with runaways the gears were invariably stripped. Holt produced his machine with link V-belt drive, at once overcoming this major problem. The belt was a 3-ply riveted leather belt, 2″ wide and tapered to fit the V-shaped pulleys. He was also first to use a single wide front wheel in a turntable for his combines, where twin wheels had been used previously. He used a hinged header from the first, so that it could be raised and lowered according to crop conditions.

The first Holt combine harvester was purchased by the partnership of Shaw, Graham, Bunch, and Lowrie in 1886 and was operated by 18 horses. It had a 14′-

The oldest known combine in existence, an 1887 Holt shown in an early Caterpillar showroom in Peoria, now on display in the Smithsonian Institution (Caterpillar Inc.)

Old Best combine with Best steamer built in 1896.
Daniel Best is seated in the buggy

cut header, V-belt drive to the cylinder, link-belt from main wheels to countershaft, and link-belt for straw carriers, grain carriers, beaters and cleaners, the rest of the drives being leather belts. It operated 46 days in its first year and averaged 25 acres a day. When this harvester was eventually sold in 1905 it had harvested 50,000 acres of grain.

In 1886 Holt had 13 combines at work, while Best had 15. At a large demonstration in 1887, with six different combine makes in the field, the Holt and Houser machines were the only ones that finished. In 1889 Holt bought Shippee, one of its largest competitors, and in 1890 no less than two-thirds of California's 2,500,000 acres of wheat were combined.

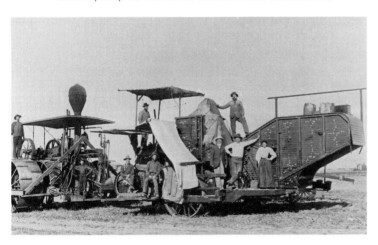

An early combine – probably a Haines-Houser

In that year the first Haines-Houser combine was built. A beautiful example of their machines—and the second oldest combine extant—now resides in the Stockton museum. This is a 1904 model built after the company was purchased by Holt in 1902, a model continued in production by them until 1912.

The oldest known combine in existence is the 1887 Holt in the Smithsonian Institution in Washington,

D.C. In 1891 Holt introduced the sidehill harvester, making three of them. The following year, due to their success, he built 20.

Many Firsts

The Holt Manufacturing Co. was incorporated in 1892. By the late 1890s, it was experimenting with auxiliary gasoline engines to replace the steam engines then in use. It became obvious that a specially designed motor was necessary, which was achieved, and in 1904 the first successful gasoline engine-powered combine was introduced and sold to Tate of Oceanside.

Other firsts for Holt had been the first combine with both wheels driving the mechanism; the widest cut harvester ever made—No. 574 with 50′ cut, built in 1893 and capable of cutting 150 acres a day; steel wheel frames on sidehill combines in 1898. The same year saw the use of the enclosed cleaner, built into the separator instead of as a separate unit, and the first use of an overshot fan.

In 1901 the straw carrier fan and double straw carrier were adopted, and in 1904 one machine was built with 60″ separator, 40″ cylinder, and 38′ cutter bar. The machine was tried first with the English method or rasp bar cylinder, but did not prove satisfactory, so it was altered to the standard toothed cylinder and concave type.

Finally in 1911 the first self-propelled production combine was introduced, and large numbers of these were built in the following years. In 1912 Holt offered nine different harvesters, from the Baby Special 10′ or 12′ cut model to the Large Standard with 18′ to 26′ cut.

It was also in 1912 that the last of the major competition was eliminated with the acquisition of the C.L. Best Co. Virtually all the other makes had been absorbed over the years. Thus was established a lineage in combine harvesters which would be inherited by Deere some twenty-four years later.

1912–1937 enter tractors and combines

The well established "Leaping Deer" implement line is joined by a growing family of tractors. Further expansion of the available products takes the company into the additional and developing line of combine harvesters, and into a quickly consolidated market position with many of its tractor models. Charles Deere Wiman takes over as president of the company in 1928.

Company Progress, Tractors Introduced

The foundations had been well and truly laid for the developments which were to take place in the fourth 25-year period of the company's history. One major item was still missing from the line to make it complete.

The branch houses and their dealers wanted a tractor to sell and had taken some steps to prove their point. Both St. Louis and Atlanta branch houses had included in their sales catalogs the Big Four "30" gas tractor built by the Gas Traction Co. of Minneapolis. This giant, hauling a 7-furrow 14″ Deere gang plow, had won the gold medal at the 1910 Winnepeg trials.

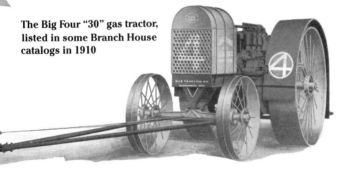

The Big Four "30" gas tractor, listed in some Branch House catalogs in 1910

It was an enormous machine with 96″ rear wheels, weighing 19,000 lbs., and equipped with a special automatic steering device shown in the illustration.

In 1912 Deere listed Twin City Model 40 tractors in its export sales specifications for Argentina and Uruguay, again for use with its tractor gang plows. The Model 40 also had a self-steering device, made by Cuddy.

Photo of Twin-City "40" tractor working in South America. These tractors appeared in Deere & Co.'s export catalog in 1912.

As a result of these pressures from the grass roots —or should it be the furrow—and the increasing threat from competition, particularly International Harvester of Chicago and Case of Racine, the board resolved on March 5, 1912, "to produce a tractor plow." C.H. Melvin was asked on July 1 of that year to build an experimental model.

Experimental Tractors

One tractor was built similar to one made by the Hackney Manufacturing Co. of St. Paul, Minnesota. This was a 3-wheeler with the single wheel behind for plowing, but reversed and in front for haulage work. It was a 3-plow tractor, but work on this experiment stopped in 1914.

Melvin tractor, with 3-furrow integral plow at work in 1912

On May 27, 1914, Joseph Dain Sr. was given the job to produce a tractor that could be sold for about $700, and by February 1915 the first tractor had been built. Again this was a 3-wheel design—apparently quite popular at that time—though in this case the single "wheel" was at the rear. It weighed 3800 lbs. and obtained over 3000 lbs. drawbar pull. By May it was decided to build three to six improved models, and by fall the second tractor was working.

It plowed 15 acres in Moline and was then sent to Minnesota, where it plowed 80 acres at a cost of 59 cents per acre, including the cost of the driver at 30 cents an hour. It pulled a 3-furrow plow with 14″ bottoms, working 6″ deep at 2½ mph. In this same field the farmer was using an 18″ Sulky single-furrow plow pulled by five horses, indicating the heaviness of the soil. In December the third tractor appeared, similar

Joseph Dain Sr., in 1910

to the second, but with a positive gear-drive transmission instead of the friction type used previously, and with chain final drive.

Extensive trials were conducted in Texas in the spring of 1916, and it was decided to build ten more tractors, in the Marseilles Works in East Moline. These were to be suitably modified in the light of experience gained. On June 13 it was reported that five tractors had been built in East Moline and one in Moline, all fitted with Waukesha engines. A new motor was being developed for these tractors by McVicker of Minneapolis to overcome the power shortage of the earlier models. The design for this was completed by July 17 and the first motors were ready by August 17.

Based on the motor costing $200, these tractors were estimated to cost about $600 to build, which would mean a sale price to the farmer of $1,200. Although this was rather higher than hoped for, it was believed the farmer might pay this for an all-wheel-drive tractor.

September 12 saw six prototypes at work, three in the Minneapolis area and one each near Minot and Fargo, North Dakota, and Huron, South Dakota. The Minot tractor had been in the area for six months. After inspection of the tractors at work, it was reported that they were the "best on the market" for the following reasons: They had all-wheel 4-chain drive, which was more durable and much quieter than gears. They could change speeds without gear clashes and without stopping even under load, thereby saving time, and they had foolproof simplicity and accessibility.

One of the tractors sent to Huron was used on the farm of John Deere dealer, F.R. Brumwell; it had plowed 110 acres, harvested 260 acres, and pulled five

wagon-loads of stone about 12 miles to Huron on two occasions. The plow used was a John Deere No. 5 Pony tractor plow with NA 214 bottoms, working 6″ to 7″ deep in heavy black loam with hard subsoil below 6″. Even when the plow was set down to 8″ or even 10″ the tractor pulled it with no engine knock or undue wheel slippage. Brumwell was enthusiastic about this tractor and believed it much better than any other he had seen. As a result he bought three tractors with the new motor, one for his farm and two for customers.

The first tractors with the Waukesha engines were satisfactory in every way except that they lacked power. When they were changed over to the McVicker designed engines, the tractors proved to do all that was required of them. The final drive chains had been beefed up to cope with the extra power. Of the four 1917 tractors with this larger engine—the balance of the order for ten—three went to Huron and one to Minot.

It was felt by both Velie and Dain that the tractors had fully justified themselves in operation, and the designation was to be 3-plow in stubble, 2-plow in

All-Wheel-Drive tractor (studio) in its final form. Ahead of its time, it had a 2-speed forward, 2-speed reverse automatic gearbox in addition to drive on all wheels.

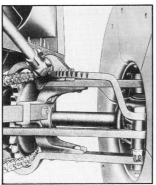

Detail on front wheel drive of the Dain tractor **Detail of the engine and its accessibility**

turf. As a result of all these trials and the reports thereon, it was resolved on September 12, 1917, to have 100 tractors built by an outside firm. This instruction was modified two months later, when the decision was taken to build them in the East Moline Works. On December 11, 1917, their construction had been provided for, and the contract for the motors issued. It was expected to complete the first 50 by June 1, 1918.

On March 14, 1918, Deere & Co. bought the Waterloo Gasoline Tractor Co. for $2,350,000.

Dain Tractor Produced

The All-Wheel-Drive tractor seen pulling a breaker plow in difficult conditions

In fact the All-Wheel-Drive tractors were completed in 1919, and all 100 went to the Huron area. They were built in the 10th Street factory in East Moline, where the first binders had been made. In its final form this tractor had a 4-cylinder motor, 4½″ × 6″ bore and stroke giving 24 belt hp and 12 drawbar hp, weighed 4600 lbs. and had an exclusive two-forward, two-reverse-speed change-on-the-move transmission. Easy worm and sector steering and other advanced features, like pistons removable without dismantling head or sump, meant the tractor was ahead of its time. It was also too expensive when compared with the Waterloo Boy—$1,700 final figure compared with $850.

The Waterloo Connection

The Waterloo Boy tractors in production in the spring of 1918 were the last style (M) of the Model "R" and the first production run of the Model "N." The Waterloo plant had its beginning back in 1892 when John Froelich, a custom threshing operator, had de-

John Froelich in 1891

cided to build a gasoline engine powered tractor to replace the steamer he was using. In South Dakota fuel for steam engines was scarce, so he placed a Van Duzen motor on a Robinson steam engine chassis. The unit worked quite well and proceeded to thresh 62,000 bushels of grain in its first season of 52 days. This is claimed to be the first tractor capable of propelling itself both forward and backward.

Because of its success, Froelich formed a company with other interested parties in Waterloo, and this company was known initially as the Waterloo Gasoline Traction Engine Co. Four further tractors were built in 1893. Two were sold but subsequently returned as they did not come up to their purchasers' expectations.

Froelich tractor on the Thompson farm, Langford, South Dakota, in 1892

In order to make the company viable it was decided to concentrate on stationary engine manufacture. Froelich's interest was only in tractors so he left the company, which changed its name to Waterloo Gasoline Engine Co. One further tractor was built in 1896 to a different design and another in 1897 and both were sold.

Twelve sizes of engines were made, from 1½ to 22 hp, but in 1906 these were redesigned and the number of models reduced to six. At the same time the name Waterloo Boy was adopted, and was to become famous both for engines and tractors. In 1912 this engine range was again modified and uprated to 2 to 14

The one tractor the Waterloo Gasoline Engine Co. built in 1896

hp and given the model name "H." These were still in production, together with a 25-hp stationary version of the tractors then being made, when Deere purchased the company.

A Mr. Parkhurst from Moline joined the company in 1911, bringing with him three tractors of his own design, these having 2-cylinder, 2-cycle engines. In 1912 he designed a 4-cylinder, 4-cycle tractor with the engine cross-mounted on the frame, and known as the Waterloo Boy Standard 25 hp tractor. It weighed 8400 lbs., the "L"-head motor had 5½″ × 6″ cylinders and the rear wheels were 62″ diameter. It had automotive type steering and the exhaust passed through the ra-

Waterloo Boy Model "H" engine advertisement showing 2-hp model

Waterloo Boy "Standard" 25-hp tractor from an early catalog

diator "chimney" to give draft cooling assisted by a centrifugal pump. The tractor was available with a 4-furrow mounted plow. Its engine was designed to run at 650 rpm, but at 500 rpm the forward speeds were 1½, 2 and 3 mph. In 1913 a Mr. Leavitt joined the company with the express purpose of designing rear crawler tracks for this model, which was then called the "Sure Grip, Never Slip."

25

Some Mystery Tractors

With smaller tractors introduced by the company certain anomalies existed and these remain unresolved. A Light tractor in 1912 appeared to be very similar to the Waterloo Foundry Co.'s "Big Chief" 8-15 model, except that as announced, it had a cooling "chimney" similar to the Standard model. In 1913 the 15 hp Model "L" was advertised with a 2-cylinder horizontally opposed 6″ × 6″ motor governed normally at 500 rpm, 48″ × 10″ rear wheels and weighing 3000 lbs.

The same advertisement quotes the Model "H" as a 25-hp tractor with 2-cylinder horizontally opposed 7″ × 7″ motor, 450 rpm, 62″ × 12″ rear wheels and weighing 6500 lbs. Yet another sales leaflet describes the Model "C," with a 2-cylinder opposed 5½″ × 6″ motor and all-wheel drive by chain including the twin fronts. These front wheels could turn through 90 degrees, allowing the tractor to turn in its own length. It was available as a 3-plow unit, although "we can attach plows in the same manner as our Model 'L' if desired," i.e. a 2-furrow mounted plow.

The above are but some of the mystery tractors advertised in 1913. The factory records do not go back far enough to determine numbers of each model built. They merely show the "L" and "LA" tractors being shipped between January 28 and June 25, 1914, again dates which are difficult to reconcile with the above advertisements.

The Waterloo Tractor Company was little different from the rest of the tractor pioneers at this time. Design was in a period of many changing ideas and concepts. Many prototypes visualized as potentially world beating tractors by their designers were conceived and built, tested and abandoned.

The Waterloo Boy Tractor

It is recorded that twenty-nine Model "L/LA" tractors were built in 1914, nine of them with 3-wheel arrangement, the balance of twenty with four wheels. All the 3-wheelers were sent to Cleveland, California, and the others were spread across the country.

With these tractors, as with the Model "R" which followed later that year, there was overlap of serial numbers between each model. When asked about this in the later context of the Waterloo Boy and Model "D" tractors, Lyle Cherry, the factory's sales manager for many years, replied that no wastage was

allowed, and if parts were available for an earlier model in sufficient quantity then additional units were assembled. This overlap of serial numbers will be detailed in one of the appendices.

In 1914 the twin-cylinder motor for the new Model "R" was designed with 5½″ × 7″ cylinders, an integral cylinder block, and head with the valves in cages. Some 116 tractors in four styles (A to D) were sold up to March 1915, the first one, No. 1026, being shipped on June 26, 1914. The motor for the subsequent styles E through L had an increased bore of 6″. The final style M of the Model "R" was fitted with a 6½″-bore motor, and this became standard for the 2-speed Model "N" and the Model "T" stationary unit.

Surprisingly, although the Standard, Catapillar, L, and C models, and presumably the H, of which no photograph or illustration has come to light, were all fitted with automotive worm-and-sector type steer-

Single speed Waterloo Boy "R" style E, No. 1,643. This style was the first with 6″ bore engine. The tractor illustrated has the optional vertical fuel tank.

Waterloo Boy "N" No. 29,583 also seen at 150th year celebrations in Waterloo

Waterloo Boy Model "R" style M plowing with 3-furrow plow in 1917

ing, all the Model "R" and the early Model "N" tractors reverted to the chain type steering used on steam traction engines.

In examining the 1906 to 1911 period when only engines were being built, it is interesting to note the growth of the company. In 1906 some 268 engines were sold; in 1907 this had increased to 1034; 1908 doubled this to 2315; 1909 more than redoubled it to 5841, while in 1910 some 13,019 were made and sold. With a factory capacity by then of 100 to 125 engines a day, it meant that up to 37,000 could be produced in 1911, some measure of the popularity of the product from the factory which Deere chose for their tractor line.

Demand for a Smaller Tractor

Before the final decision had been taken on which 3-plow tractor should be built and marketed, it had become evident that two other lines of thought must be investigated. These were the question of a smaller tractor for the thousands of small farmers across the country, and a mechanical means of cultivating the increasing corn and bean areas of the nation.

Concurrent with Dain's development, Max Sklovsky, head of Deere's design department, was building a prototype 2-plow tractor of similar 3-wheel arrangement, called the Model A-2 and built in the Marseilles plant. It was without doubt the first tractor with a one-piece cast iron body which included the engine sump, giving simple, strong construction. A Northway

4-cylinder motor was used, and on November 20, 1915, it was field tested. It worked well except that it was impossible to steer, as its front wheel drive had no differential.

It was therefore rebuilt with a pivot axle with automobile-type steering and called the Model B-2. Starting in July 1916 it operated for eight weeks continuously and successfully. At an estimated sale price of $900 it would have been acceptable, but the problem with a tractor of this size was that the 4-cylinder motor, running at 1000 rpm, was not suitable for burning kerosene.

Again one further design, to be known as the Model D-2 and to sell at about $600, was produced with a single-cylinder motor and with other modifications suggested in the light of experience with the previous tractor. The war intervened, bringing its development to an end.

Sklovsky "B-2" experimental tractor, with Max Sklovsky at the wheel

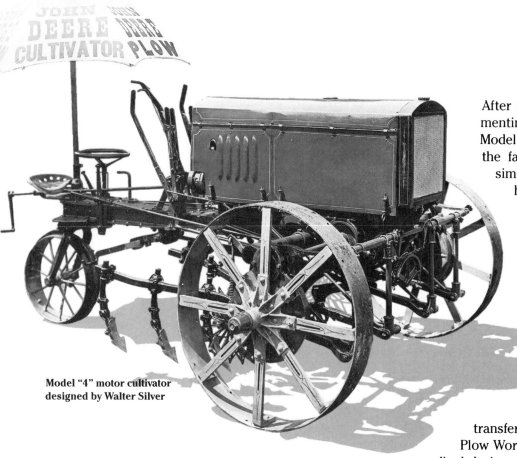

Model "4" motor cultivator
designed by Walter Silver

After some twelve months' experimenting and modifying, first on a Model 3 and finally a Model 4, with the family likeness of the Dain—similar front wheels, radiator and hood styling—the prototype was ready and five were built. They were all working by June 1918, disking, planting, cultivating, drag harrowing with 20' drags, and binding. Since square turns could be made there was no stopping at corners, and the unit's performance was trouble free. The only drawback was again lack of power.

Development work was transferred from East Moline to the Plow Works the following year and the final design, again for a one-row machine but like the 2-row, was built in 1921. Due to the disastrous farming market at that time it was decided to discontinue the program. These early motor cultivators led later in the twenties to the general-purpose tricycle-type tractors which became the most popular in the States for several decades.

Motor Cultivators

The second investigation during the 1916-1917 period was into "motor cultivators." Joe Dain Sr. produced the first ideas and these were further developed by Theo Brown. The first unit appeared with a 7½-hp New Way air-cooled motor, but once again this was not strong enough, so McVicker, designer of the Dain tractor's engine, designed a 2-cylinder hopper-type water-cooled unit, and this was built by the Amanco Co.

When cultivating, the unit was used with rear wheels driving, but for general-purpose use this could be reversed like the original Melvin tractor. In September 1916 it was tried successfully with a mower; that fall a complete redesign took place, and that winter, the Marseilles plant built 25. They were one-row units called "Tractivators," and were sent to most Midwest branches.

The appearance on the market of a 2-row outfit from the International Harvester Co. ruined the prospects of Deere's machines, and the single-row idea was dropped. In June 1917, Walter Silver of the Deere design team developed a 2-row model with an Avery motor. It was a 3-wheeler with rear wheel steering.

The Waterloo Boy Development

Even before Deere purchased the Waterloo plant, the Waterloo Gasoline Engine Co. had been experimenting with a totally enclosed-gear tractor, built very much like the Wallis Cub and known as the "bath tub." It did not prove satisfactory on test, the problem being with the final drive, as Wallis experienced with their model, this despite various options being tried.

The experiments continued under the new ownership, but in the meantime the Waterloo Boy Model "N" was improved, being fitted with automotive-type steering from No. 20,834 on and with a riveted instead of a bolted frame from No. 28,094 on.

The answer to the final drive problem was found in 1918 by using roller chain for this purpose, doubtless due to the success of chain drives on the All-Wheel-Drive tractors.

Improved 1920 Waterloo Boy "N" with automotive steering and riveted frame. This model has the higher fuel tank position and the combined gasoline and oiler tank.

Overhead view of same tractor

Advertisement of early Waterloo Boy "N" 2-speed model from the Farm Implement News

Late Waterloo Boy "N" working with Moline, Ill., city road grader in January 1921

Experimental Waterloo Boy tractor
shown in July 1919

The Famous Model "D"

From 1919 to 1922 several versions of the modified Waterloo Boy tractors were built in four styles. Seven of Style A were numbered 100–106, seven of Style B 200–206, and twelve of Style C 300–311, with No. 301 being rebuilt and No. 305 scrapped. Finally style D, the one adopted for production, was originally No. 401, but became No. 30,401. The tractor model which stayed in production longer than any other, before or since, had arrived.

During the twenties the company's largest export market was the Argentine, but by a happy coincidence at the low point of the Great Depression, Russia adopted its first 5-year plan. As a result of this, large quantities of Model "D" tractors and 3-furrow plows were exported to them. Chris Michaelson, boxing foreman at the Plow Works, recalled the colossal job of boxing these plows for export. They took one-half of the old shipping platform.

It is interesting to compare the number of Model "D" tractors sold to the Argentine and to Russia over a 6-year period.

The same tractor plowing with a 3-furrow No. 5 plow

Pre "D" during summer of 1921 with Waterloo Boy round spoke rear wheels, left hand steering position and two-section steering rod

Rear view of the same tractor and plow. The right hand Waterloo Boy driving position is shown up well in this picture.

Pre-production Model "D" in 1922 with jointed steering rod

Year	Argentine	Russia
1927	15	31
1928	300	51
1929	2194	2232
1930	700	4181
1931	0	1679
1932	0	31

Three specially equipped Model "D" tractors on dual rubber tires at the factory in Waterloo in July 1931

One of the later experimental Model "D" tractors being evaluated near Waterloo, Iowa in 1922

One of the first 50 Model "D" tractors with horse-drawn binder in 1923

A very early 1923 production Model "D" with 26″ spoke flywheel, and cultivator

Model "D" assembly lines in August 1924

31

A Smaller Tractor Again Needed

Once the Model "D" was in full production and proving an outstanding success, thoughts once again turned to a smaller tractor. In 1926 an experimental tractor design was developed and at least five machines were produced. Nos. 1 to 4 were built, with No. 1 being of the tricycle type. The records then show a gap to tractor No. 9, again with the tricycle configuration but with wide rear tread.

First Tractor With Mechanical Lift

Production of the Model "C" began in earnest on March 11, 1927, with serial number 101, and 25 tractors were built to number 125 by May 5. Sixteen of these first 25 were recalled and rebuilt.

The next batch of tractors were numbered 200,001 to 200,069, and of these 37 were rebuilt and renumbered. There is then a gap in the records to number 200,080, another gap to 200,099, and finally numbers

Model "C" No. 200,109 at the Waterloo, Iowa Show, nicely restored

Early Model "C" (10-20 hp) and integral planter in May 1927. Note unusual front axle, left hand steering rod, and no name on radiator top tank.

Another early "C," fitted with a D-type exhaust muffler and vertical air intake and a 5-row front-mounted cultivator.

200,102/3/6/7/8/9/10 were built. New serial numbers for the rebuilt tractors commenced with 200,111 and were interspersed with a further 24 new and original machines ending with number 200,202. Another gap in the records follows, with the production "GP"s, as they were subsequently known, starting at 200,211. The oldest "GP" was rebuilt "C" No. 200,121, which in turn was originally No. 104 of the first batch. For a full listing of these early "C"s consult the appropriate appendix.

Over the eight years of its production the Model "GP" underwent several changes. The engine size was increased from 5.75″ × 6″ to 6″ × 6″ from serial

Tricycle Model "C" and cultivator, June 1928

number 223,803 on. An orchard version was introduced in 1930 and numbered separately starting from 15,000, and a few of these were purchased by the Lindeman Company in Washington State and fitted with crawler tracks. At the time of this writing, 11 of these latter are known to exist in the hands of collectors.

The Tricycle Row-Crop Tractor

The only complete Model "GP Tricycle" known to exist, No. 204,213

One further option had appeared over the "GP" production period. Just as the first Model "C" had been a "3-wheel" type, so during 1928-29 23 "GP"s were built as tricycle-type tractors. At least two of these were built with 68″ wheel tread and anticipated the production of the Series "P."

View of Series "P" from sales literature in 1930

In 1929 the "GP Wide Tread" went into production, and again it had several updates during its lifetime, the last and most fundamental being the change from sidesteer to overhead steer. At the same time it was redesigned to give a better view for the operator, with a slimmer hood and fuel tank, allowing for improved cultivating performance. With only 445 of this model produced, they are currently valued more highly than most of the other tractors built in the 1930s.

The "P" was a special series of wide-tread tractors with a narrower rear axle for work in the potato fields

"GPWT" with experimental overhead steering and rear dished "potato" wheels, July 1931

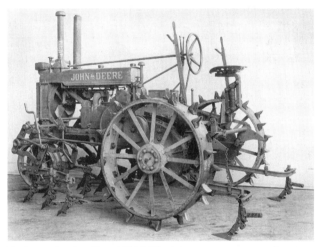

Early production Model "GPWT" with overhead steering fitted with GP 232 2-row cultivator, August 1932

of Maine, to which state all were shipped. Some 203 tractors were allocated "P" numbers, starting with P-5000, and of these 150 were built new and 53 were rebuilt "GP"s with serial numbers between 222,219 and 222,395. As No. 5183 went to the experimental department, 202 tractors were actually sold.

The Model "A"— The First Tractor With Hydraulics

These late wide-tread tractors gave an indication of the next development and the resulting 2-cylinder tractors produced, the Model "A." Before its introduction for the 1934 season, at least eight units were built, two with a 3-speed transmission like their predecessors and six with 4-speed, and known respectively as Models "AA-3" and "AA-1." They were built in late spring 1933 and numbered 410,000 up. All but the second were rebuilt and given new numbers. A further four tractors were made and subsequently rebuilt and also given new numbers, after which the main production of Model "A"s began in April 1934 with serial number 410,012. The second production tractor appears each fall at the Mount Pleasant show in Iowa, but it is interesting to speculate on the possibility of the existence of any of the first eleven tractors which were given new numbers. These are shown in the appendices.

Model "AA-1" tractor and GPA 462 cultivator on December 15, 1933. Note the lack of name on the rear axle housing.

Model "AA-1" tractor and GPA-262 cultivator the next day, showing bail-type air cleaner intake and center fuel tank cap

An example of the "A" with open fanshaft, No. 411,880, at the Sully, Iowa, sale in July 1987

Rubber Tires

We have already seen that some Model "D" tractors used for industrial purposes had appeared on rubber tires as early as 1931. The first half of the thirties saw these experiments extended to the agricultural scene, and to tractors and equipment supplied for other uses.

Model "109," the "GPO" on rubber tires for golf courses, 1933

A good example of these special applications was the "GPO" model fitted with low-pressure balloon tires. This unit was offered to golf course owners as the Model "109," and its low profile and light ground pressure were cited as ideal for fairway upkeep.

Farming on Rubber

The most impressive result of rubber tire use on tractors was the great increase in work output. The higher field travel speeds made more work output (horsepower) available to the field implement. While the implements required redesign to accommodate the higher working speeds, this was beneficial in sev-

AT LEFT—Rear view of the new John Deere Model A with pneumatic tires cultivating two rows at a time. Rear rigs can be furnished for cultivating out the tracks left by the tires.

AT RIGHT—Side view of Model A cultivating corn. Pneumatic tires have an advantage in cultivating because they pick up less dust than do steel wheels.

Two views of the Model "AA" on rubber tires cultivating two rows of corn

eral ways. Tractor manufacturers found the development of rubber tires permitted substantially more powerful engines in the same tractor chassis. Implement manufacturers in adapting for higher working speeds also used much the same factory machinery to provide the increased benefits to the farmer.

The use of rubber tires was one of the most significant developments affecting farm machinery history. It enabled tractor cost per power unit (hp) to remain essentially constant for many years of business economic inflation, and greatly contributed to enabling the U.S. farmer to become the world's low-cost agricultural producer.

In farming, all the tractor models were available with low-pressure pneumatic tires, on wheels supplied by French and Hecht. Initially these were all of the round-spoke type. An obvious example for which rubber tires were ideal was grassland because they gave greatly reduced damage to the ground surface. Mowing, raking, sweeping, and loading hay were among the first operations to benefit from this development.

When it came to tillage the chief problem was one of traction. Accordingly, rear wheel weights were offered. Once this snag had been realized and largely overcome, the advantages of operator comfort from reduction of vibration and shock were appreciated and advertised.

Road Use

Use on the public highway meant that demand for higher road speeds soon followed. In addition those implements most often transported along the country lanes were other candidates for rubber tire equipment. Gradually the whole farm implement line was to be so fitted if required. The beginnings of a speed-up in farming operations and techniques had arrived.

The Model "B" Introduced

With the gradual increase in horsepower, the problem which had arisen regularly over the previous twenty years again appeared—the requirement for a smaller tractor for the small farmer and as a backup on larger farms.

Therefore, one year after the advent of the "A," the "B" was born. First advertised as a garden tractor, it soon proved to have much wider applications. Similar in every way to the larger model, it was designed to replace four horses, compared with six for the "A," and both had

Early Model "B" advertisement, showing tractor without John Deere name cast on radiator top tank

Early Model "B" and B-113 cultivator in December 1934 with center fuel tank filler cap and 4-bolt pedestal

Experimental Model "B" (later serial numbered 1803) with no "John Deere" on either the radiator or the rear axle housing, on the Administrative Center display floor, July 1987

even greater daily output potential than this suggested.

Late in 1935 both these models were offered with single front wheel equipment, followed shortly by a wide adjustable front axle option. At the same time two further options were introduced—a standard (or regular) 4-wheel model, the "AR" and "BR," and for orchard or grove work the "AO" and "BO," similar to the standard models but built low down and with independent rear-wheel brakes as standard equipment.

Industrial Applications

Model "DI" and Caterpillar road grader in 1935

It had been appreciated from early days that agricultural tractors were capable of use in many industrial applications. Photographs show the Waterloo Boy grading rural roads, and from the commencement of its production the Model "D" was available as an industrial unit, later called the "DI." With the announcement of the "A" and "B" series of standard tractors, Models "AI" and "BI" were also added to the

Model "AI" pulling rail wagon in the Waterloo Works in July 1936

industrial types on offer. Similar to the standard tractors, they were offered with attaching pads at the front of the frame, independent rear wheel brakes as on the orchard models, and armchair-type seats.

An additional side power takeoff installation was offered for use with cranes and hoists. Equipment for industrial use included a street flusher and sprinkler, a Van Brunt road maintainer, trailers, La Plant-Choate cranes, Parsons snow plows, Hyster hoists, and Detroit sweeper and snow brush. Truly the beginnings of an industrial line.

Model "BI" equipped with electric lighting with road washer. Note the drilled and strengthened frame to allow the attachment of equipment.

Stationary Engines

Supporting these tractor developments, the company still offered its range of stationary engines. The Model "E" enclosed-crank engines in 1½-, 3-, and 6-hp sizes had replaced the Model "H" series of Waterloo Boy engines in the twenties.

The Model "T" 25-hp unit was replaced first by a spoke-flywheel type Model "W" in 1925 and 1926. This

in turn gave way to the solid-flywheel types "W" and "WSP," the latter for special pumping operations and supplied without radiator or fuel tank.

By the mid-thirties, the "W" series consisted of the standard W-111 35-hp model, the W-113 which had replaced the "WSP," and the "W" cotton gin version with a special protective radiator screen.

Root & Vandervoort "New Triumph" engine and washing machine advertisement

Model "E" and pump jack from 1927 sales brochure

Model "W-111" No. 4,938 and "W-113" No. 3,924 at Waterloo Show in July 1987

A Baby Tractor for the Line

Faithful reproduction of Model "Y" tractor at Waterloo show by Jack Kreeger of Omaha, Nebraska

When the company stopped production of the "GP" and "GPO," this left eleven models available during 1936. In that year one other development was taking place. The Wagon Works built twenty-four small Model "Y" tractors for garden or horticultural use. Weighing only 1340 lbs., they were offered in the 1936 Wagon Works sales list at $532.50 and a Y-636 garden planter and Y-630 garden cultivator were also listed. The tractors were fitted with a Novo 2-cylinder vertical engine, a Ford Model A transmission mounted on the rear of the clutch housing, and a Ford Model A steering gear with no support for the steering wheel. The welded frame was different from that of the subsequent Model "L."

As these tractors were looked on as experimental machines, they were all recalled, but recently a reconstruction of one based on a Novo engine which had been discovered was shown at the 150th year celebrations in Waterloo in 1987.

Studio picture of Model "Y" tractor in 1936

Following the success of the idea, an uprated tractor, the Model 62, was introduced in the summer of 1937, seventy-nine being built. These were followed that fall by the unstyled Model "L." The two models are distinguished by the JD symbol on the casting below the radiator and another on the rear axle banjo of the 62, these places being left unadorned on the later model. Other detail differences in the fenders and rear wheels also help identification.

The engine used in both models was a John Deere of Hercules design, made by Hercules at their works in Ohio. It was a vertical 2-cylinder gasoline engine of 8 hp, and the tractors were designed to operate a single furrow 12″ plow, a one-row corn cultivator, or a 5′ disk harrow. A belt pulley was available, but no PTO shaft. The weight had increased a little over the "Y" to 1630 lbs., and all these models had 3-speed gearboxes. These brought to 12 the tractor models available as the company reached its hundredth birthday.

Rear view of Model "62" showing JD on rear axle housing, tractor No. 621,048

Model "62" tractor with distinctive JD on casting below the radiator

Tillage and Cultivation Equipment

Tillage and cultivation equipment for horse-drawn implements had changed little since the beginning of the century. Tractor-drawn units had similarly remained the same for many years. Attached to the tractor drawbar, the plows, disk harrows, disk tillers, field cultivators, various types of drag harrows, and pulverizers had

William Butterworth

varied only in size to match the increasing power of the tractor, whether steam or internal combustion.

With the advent of the general-purpose tractor, with first mechanical lift and then hydraulic, all that changed. It was now possible, in fact desirable, to mount much of the tillage equipment on the tractor itself. The mechanical lift on the Model "C" was a world first. Cultivators up to 5-row were experimented with, and speed of operation over the one- and 2-row horse machines showed an enormous increase.

Model "D" tractor and disk tiller fitted with rope-controlled power lift for raising and lowering disks

Plows

At the same time the number of models of each type of machine increased. A special light plow was introduced for use with the ubiquitous Fordson tractor, while light and heavier versions of the single- and 2-furrow plows were offered; but in the 3- to 5-bottom class, the Nos. 5 and 6 plows had remained virtually unchanged since Waterloo Boy days.

With the demise of the steam engine, the engine gang plows had disappeared from the sales catalogs over the 25-year period, while the original Jumbo

Model "D" tractor with No. 2 two-way tractor gang plow, at work in September 1928, making an excellent job of burying a tall crop residue

Waterloo Boy tractor with Pony double-action disks and Dunham Culti-Packer at a large demonstration in July 1918

Grub Breaker had become the new Models 11 and 12. Two-way tractor plows, both drawn and integral, had been added to the line, as had listing plows, bedders, and middlebreakers. The 20 series disk plows of 3 to 5 disks had given way to the smaller 92 and 93 2- and 3-disk models, and the larger 100 series with up to 6 disks. Disk tillers had been introduced, and were becoming increasingly popular with wheatland farmers of the United States and Canada. Disk harrows had multiplied in type to suit all sorts of farming conditions. From the Model "B" single-action and Pony double-action disks had developed no less than eight different models, single and double action, in widths up to 21'. The Model "S" single action became famous as the outstanding value for the larger farm, with its capacity of 100 acres a day.

Cultivation Equipment

Spike-tooth harrows were available up to 30' width (six sections), while spring-tooth harrows had become popular for weed control. The "CC" field and orchard cultivators from the Van Brunt factory had become as famous in their sphere as the "S" disks. A model "N" with wheels outside the frame was also offered. Deep-tillage cultivators and rod weeders

Twelve-foot tractor-drawn rod weeder used extensively in summerfallow areas

rounded out the tillage line. As a pre-tillage tool, rolling stalk cutters, which had been in use since 1879, were still available, heavied up to cope with tractor power. Machines of this type had to take heavy punishment when dealing with corn stalks and cotton waste, so they were equipped with extra-long gray-iron bearings.

John Deere spring-tooth field cultivator with nine tines showing ease of depth control from the tractor seat

41

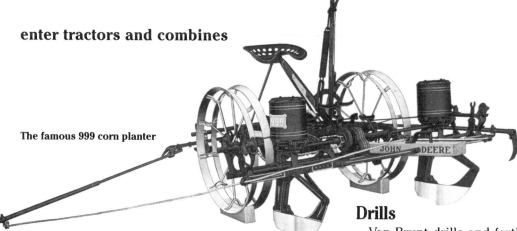

The famous 999 corn planter

Planters

Corn planters had not changed very much. The most famous of all, the 999, developed in 1913 from its equally illustrious forebear, the No. 9, was still in production in 1937 at its home plant, the Deere & Mansur works. A 4-row version, the 450, had been developed for use with the "A" tractors, and a 6-row beet and bean planter had been added to the line, but the individual planter units were little changed. "Planet Jr." vegetable planter units could be made up into 2-, 4-, or 6-row B-636 tractor vegetable planters, and the famous Hoover line of potato planters was available for one or two rows.

Hoover 2-row potato planter with fertilizer attachment and front truck

Drills

Van Brunt drills and fertilizer distributors continued to dominate the seeding of small grains and chemicals. The drills came in many shapes and sizes, from the "EE" all-steel fluted-feed to the "SS" double-run models of similar overall construction. The "LL" press drills and "F" combined grain and fertilizer drills added to the types available. The early grass seed drills had become the new "OO" rust-resisting galvanized steel box model which was to stay in the line for many years. Drill spacing was available from

Model "EE" grain drill with 5250 lbs of weights on its box illustrating its strength, pictured in 1933

the 4″ of the grass drill to 16″ on the press drill, depending on requirements. The "SS" was available as a semi-deep furrow drill, and both the "EE" and "SS" for deep furrow work. Finally the "Z" lister drill rounded out the models, and was used in winter wheat areas, where the problem was to prevent winter killing, blowing and heaving.

An early "D" tractor and disk harrows with mounted Van Brunt seed drill, in 1929

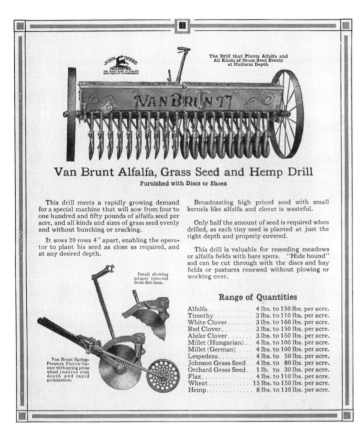

Advertisement for early Van Brunt alfalfa, grass, and hemp seed drill, 1926

Cultivators

Tractor power had cut the costs of cultivating, from the days when one man and a horse-drawn one-row cultivator needed 37½ days to cultivate 100 acres three times over. In the company's centenary year, one man with an "A" tractor and 4-row cultivator could complete the same work in six days. The "A" had been the first tractor with hydraulic lift, just as its predecessor "GP" had been the first with mechanical lift, and these aids had further speeded up the operation of mounted equipment at row ends. Front, mid-, and rear-mounted toolbars were all available for different job requirements. The mechanization of America's staple crops—corn, beans, and cotton—had been mastered.

Haymaking Equipment

The John Deere way of making hay had long been a byword with farmers. It was recognized by leading hay specialists as the most economical and practical

Advertisement for No. 1 and 2 mowers, 1927

method of making high-quality hay. This involved moving the mown swath fairly soon after cutting, to prevent the leaves from drying out and shattering before the stems could release their greater moisture content. The key to this system was the side-delivery rake with floating cylinder, introduced from the Dain plant in 1914. In addition to its unique cylinder it had a universal-joint drive, which gave positive uniform flow of power to the rake at all times, and curved teeth which lifted the hay and ensured a fluffy and airy windrow. Curved teeth are as essential to a rake as to a hay fork.

A Dain side-delivery rake pulled by two horses in a hayfield in July 1920

43

Mowers

Model "GP" tractor with mounted power mower and side-delivery rake in July 1928

First, of course, one has to mow the grass. The original Dain vertical-lift mower had given way first to the John Deere enclosed-gear high-lift mower with 5' or 6' cut and then to the No. 1 Regular Frame 3½' to 5' cut and its larger companion, the No. 2 Big Frame 3½' to 7' cut machine. These were followed in turn by the Nos. 3 and 4, respectively, the latter becoming the Big 4, which stayed in production well into the 1950s. Finally the No. 5 power mower was introduced, positively driven from the tractor's PTO and again available up to 7' cut. This was a semimounted machine with a single rear wheel, initially steel but later with rubber tire equipment. It acquired an excellent reputation worldwide.

Transporting the Hay

Model A306 Raker Bar-Cylinder hay loader at work in June 1936 behind two horses and an old 4-wheel wagon. Some hard work was required of the man loading.

Once the hay was cured, the age-old method of loading it onto wagons, using the Dain rake-bar loader or the further-developed John Deere single-cylinder and double-cylinder loaders, was one way of moving the hay from the field. Dain had introduced the sweep and stacker method in the previous century, and no less than seven different sweeps and a similar number of stackers were offered.

Four different Dain power-lift sweep rakes from a 1922 catalog

These had been reduced to a standard model sweep rake, available with operation from the tractor's lift, and three stackers of different capacity, chiefly because of the introduction of the windrow pickup press. Gathering the hay directly from the windrow made with the side-delivery rake eliminated the sweeping operation, thus saving valuable leaves.

Presses

Stationary presses of various sizes, horsepowered or tractor belt driven, had been in use throughout these 25 years and before that, and were to continue to be available for quite a time afterward. The big

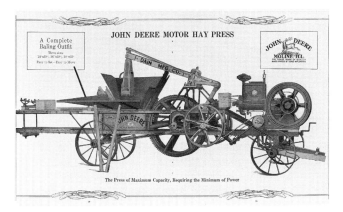

A 1915 advertisement for Motor Hay Press with Root & Vandervoort engine drive

step forward came in 1936 with the introduction of the new windrow pickup press. At one stroke all the operations from the windrow to the hay in storable form were reduced to one process.

In average crop conditions it required two men to operate the machine, but in heavy hay crops, particularly in Europe, it was not uncommon to see four men on the baler—one controlling the hay input, one feeding the wires through the grooved blocks used, one returning those wires, and the fourth man tying them.

The early models were equipped with Lycoming motors, but these were replaced later by Hercules engines. The steel wheels of the originals were replaced by a rubber tire option, but wartime restrictions would require a return to the former.

Harvesting

Harvesting was where other major developments had taken place. Apart from two corn cutters made in the Dain factory, one a 3-wheeler with steel frame and the other a 4-wheel wooden-frame machine, the harvesting equipment in the line was a grain binder and a corn binder. These machines were basically designed for use with horses, but conversions for tractors were soon available. It was the late twenties before the John Deere 10-foot tractor binder was introduced, with PTO drive and with extra strength to stand up to continuous tractor work. In 1929 a rice version was added, with either 7′ or 8′ cut.

Model "D" tractor and power grain binder with remote controls for operation by one man, in a wheat crop in August 1930

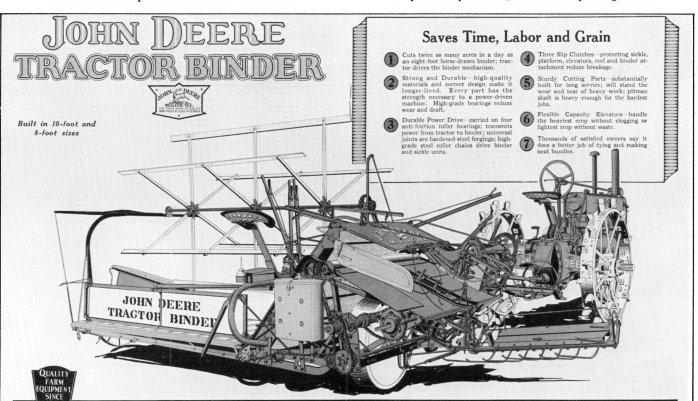

Advertisement for tractor grain binder with Model "A" tractor in March 1937

The Corn Picker

Horse-drawn corn picker operating in the 1927 harvest

Model "GP" tractor with 10 one-row corn picker in 1930.
Note the shields for the chain drives.

In the early twenties, however, it was becoming increasingly evident that two other harvesting machines must be considered, the combine harvester and the corn picker. Under the guiding hand of the company's president, Charles Wiman, who was keen to introduce new products into the line, experiments were instituted and 23 prototype corn pickers were made for the 1925 harvest season.

Although the year was a difficult one for harvesting, the machines proved successful, and after another trial in the 1926 harvest, the new corn picker was put into production the following year. For the 1928 selling season no less than five models were offered – the No. 1 horse-drawn, No. 2 tractor-drawn, No. 3 tractor-operated from the PTO, No. 4 horse-drawn with a Ford motor drive, and the No. 5 tractor-drawn with the same motor.

1929 saw the introduction of the 10 one-row corn picker and the 20 2-row model, both of which remained in the line until the centenary year, and in the case of the 10 for much longer.

The Combine Harvester

The second pressing problem was the question of the combine harvester. Again Charles Wiman was keen to see these added to products offered by Deere. The first thoughts were to purchase an existing company, as had been done so successfully on previous occasions with other products. The Universal Harvester Co. of Kansas City, already making combines, was considered, but wiser counsel prevailed. The company purchased one of the Universal combines instead, another one from Massey Harris, and built a prototype in the Harvester Works for the 1925 season.

Model "D" tractor and No. 1 combine unloading in the 1927 harvest in an Illinois wheat field

Pre-production combine harvester in the 1926 harvest pulled by a Model "D" tractor, and with wagon alongside for the grain. The reel is gear driven and the combine is driven by a 4-cycle motor.

Combine Production Commences

It compared well with the Massey machine and was much better than the Universal, so after further tests in 1926 the larger of two machines, the No. 2, was made available in some quantity to dealers for the 1927 harvest, followed the next year by the smaller No. 1. This machine was available with 8', 10', or 12' cut, compared with the 12' or 16' cut of the No. 2. Both machines were driven by a 4-cylinder Hercules motor, and were sturdily built. The demand was for somewhat lighter construction, so in 1929 the No. 3 joined the line to replace the No. 2.

No. 3 windrow model combine with Model "D" tractor in the 1929 harvest

A Sidehill Model

The No. 4 sidehill combine was built experimentally, but just at this time Caterpillar offered Deere their Western Harvester division, builders of the celebrated Holt combine line. Consideration of this offer temporarily halted further experiments on sidehill machines. The asking price proved to be too high, and negotiations were temporarily halted.

Deere's first sidehill combine, the No. 4, at work in the 1928 harvest. This model did not go into production.

A Smaller Combine

The No. 5 model replaced the No. 1 in 1929, again being of lighter construction. A still smaller combine was envisaged, to become in due course the No. 6, but before this was ready for the market the No. 7 (8' cut) machine appeared in 1932, and two years later

A No. 7 engine-driven, 8'-cut tanker combine with weed seed cleaner attachment, harvesting soybeans

the No. 5A arrived as an updated version of the No. 5. The year 1932 had also seen replacement of the No. 3 with the No. 17.

When the No. 6 finally arrived on the market in 1936 a full range of combines was available from the company: the 6'-cut No. 6 with either PTO or engine drive, the 8'-cut No. 7, the No. 5A with 10' or 12' cut, and the 17, largest of the line, with 12' or 16' cut. All four models could be purchased with grain tank as standard, or with sacker attachment as optional equipment. Other extras available were straw spreaders, windrow pickup attachments, and various items of special equipment such as different types of sieves and different sizes of pulleys and sprockets. Grain lifters for laid crops were another essential extra available for all models.

Caterpillar Line Acquired

To this comprehensive range was added the Caterpillar-Holt line in 1936. In 1935 Caterpillar had made renewed overtures to Deere. They were hoping to come to a reciprocal arrangement on tractors, since Deere did not wish to make crawlers at this time, and Caterpillar did not want to make the wheel tractors which their dealers were asking for to complete their line. Since Caterpillar already had many dealers and distributors in overseas markets, this represented an excellent opportunity for Deere to expand their outlets.

Coinciding with this new arrangement between these two major concerns, Caterpillar also decided to concentrate entirely on industrial products. This time they offered to give Deere without charge their existing stock of Model 36 hillside combines, together with all the relevant spare parts, working drawings, patents, templates, dies and jigs, and all the other works equipment required by Deere to continue their manufacture.

Caterpillar had been formed from the merging of Holt and the second Best company in 1925, so a rich heritage of combine history dating back to the 1880s suddenly became Deere's.

The Model 36— Longest-Production Combine

Three 36 combines at work

The Model 36 had been designed and built first by Holt, and continued in production by Caterpillar, which had added their Models 34 and 38. As Deere had competing models with these, they were not included in the deal, Caterpillar agreeing to sell the remaining stock machines and to continue to service them and stock parts for them as long as required.

The line of four John Deere combines thus became five with addition of the 12'-, 16½'-, and 20'-cut level land, medium hillside, and extreme hillside machines. To the Model 36, Deere added the Model 35 Hillside 12' and 14' cut in 1937, and a smaller Model 33 Hillside in 1940. Even this large range of combines was to be added to soon after the company's second century began.

A Thresher Added to the Line

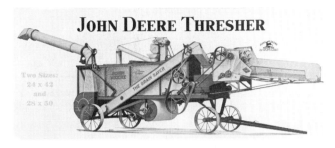

The Grain Saver thresher, John Deere's first model, in 1929

One other "harvesting" machine must be mentioned at this time. The company had surprised everyone in 1929 by adding a thresher to their expanding line. The Wagner-Langemo Co. of Minneapolis had run into hard times, and offered Deere 23 complete Grain Saver machines plus all their factory tooling for $42,500.

The John Deere thresher had become a reality. From its introduction it was made in two sizes. The Grain Saver was built as $24'' \times 42''$ and $28'' \times 50''$, and could be distinguished from its successors because its straw blower and fan were alongside the machine rather than at the rear. The machine was redesigned with rear-position straw blower, but with the same size separators as above, and with the rack-type straw separation unit of the original design. Late in 1937 it was again redesigned with straw walkers and other modern features, and the two sizes became $22'' \times 36''$

and $28'' \times 46''$, the straw walker separation having eliminated the necessity for the extra width in the threshing unit.

A Cotton Harvester

With the introduction of the first production cotton stripper, following a year or two of experimenting and with the further addition of a grain windrower, the company's harvesting line was nearing completion.

A grain shock sweep had been introduced for the "A" tractor, and this was improved and made available for the "AN" and "G" models late in 1937.

The year 1936 had seen the addition of the 15 one-row and 25 2-row push-type corn pickers, also for use with the "A" and "G" tractors. These units gave to corn picking the advantages later given to grain harvesting when self-propelled combines appeared.

Experimental cotton picker photographed in November 1928

Redesigned thresher of 1931

49

Hoover potato digger with
Ford engine drive, single-roller front truck
and digger extension, from a 1927 catalog

Root Harvesters

Root harvesting had begun many years earlier with
the introduction of Hoover potato machines to the
line. First single-row and subsequently 2-row potato
diggers had been introduced in 1919 and 1929 respectively, but it was 1937 before the company offered an early and very simple type of beet lifter, the
33M mounted type and the 33T trailer version.

Hoover potato digger at work pulled by four horses

Barnyard and Transport Machines

Of miscellaneous machines there were a multitude
offered over the last 25 years of the company's first
hundred years in being. The barnyard and maintenance category included manure spreaders, most no-
tably the famous Model "E," "the spreader with the
beater on the axle," which was still available in 1937,
along with a PTO-driven model.

The original low-down spreader with power drive pictured in
March 1931

Wagons, trucks, and trailers in different types met
all farm needs, from the original Western and Fort
Smith farm wagons and various 4-wheel truck and
wagon chassis to the buggies and surreys built in the
Reliance works in St. Louis. The wagons were still
available in 1937 for horse-drawn requirements, but
to these had been added a range of tractor-drawn
units, the 951 and 902 series trailers, the 20 and 25 2-
wheel trailer carts, and the 802 all-steel gear. All of
these could be purchased on rubber tires and with
Timken roller bearing wheels.

Corn Shellers

An experimental Model "D" tractor with an early corn sheller in 1922

Corn shellers, both hand-operated and tractor or engine-driven, had been upgraded over the years from wooden construction to all-steel, and had even been mounted on motor trucks. Models 4-A, 5, 9-A, and 10 were the latest examples of these popular machines.

The 10 cylinder-type corn sheller equipped with universal swivel feeder and raising device, wagon-box elevator and spouts, heavy running gear, and tractor hitch as shown in 1933 catalog

Elevators

Elevators too, of both the portable and inside-cup grain types, had changed in their construction from wood to steel, and tubular small grain units were latterly marketed up to 33′ long. From horse power to engine or tractor drive was an inevitable progression.

These major units were in the line continuously over the twenty-five years covered in this chapter, but many other items were added and on occasion ceased to be. Feed mills, hammer mills, grain cleaners and graders, concrete mixers, wheel barrows, yard scrapers, bag trucks, Letz mills and grinders, cream

A 1927 picture of a portable grain elevator, bridge-trussed and non-rusting, with two horse-power, triple-geared, No. 255 drive

separators, weighing machines, Papec silage cutters, sprayers, ensilage blowers, and hand tools, still made in the tradition of the master blacksmith, gave the company as full a line as any in the farm machinery business. The stage was set for growth, in its second century, into the largest supplier of farm equipment in the world.

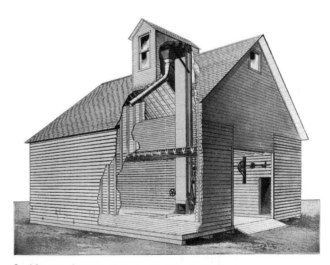

Inside cup elevator, showing positioning in barn, from a 1927 catalog

1937–1947 styling and hydraulics

The company engages an independent stylist, Henry Dreyfuss, to create a new image for the tractor and combine line. New products are held back by wartime constraints. Small tractors developed. Electric starting and implement hydraulic controls become standard. Self-propelled combine harvesters are introduced. Charles Deere Wiman continues his leadership as chairman.

Charles Deere Wiman, president from 1928 to 1942 and 1944 to 1955. Burton F. Peek was president while Wiman was in the U.S. Army.

Deere's Second Century

The second century of the company started on a note of achievement. There were twelve tractor models in the line, and the combine market was fully covered, from the small 6'-cut PTO-driven machine to the giant hillside and the level-land 20'-cut models acquired with the Holt heritage. Products in the tillage, cultivation, planting, and haymaking spheres gave as full a cover as any other company's line.

Truly this represented a cause for satisfaction. The final product was simple, reliable, long lasting and easy for the farmer of the thirties to service and maintain. Increasingly the competition was attempting to improve the appearance of machines by developing some form of styling in order to compete with Deere's strong suit, simplicity.

Waterloo Tractor Works in August 1939

Model "AN" 4-speed styled tractor on
rubber tires—studio picture, July 1939

Styling the Line

It was decided that the whole line should be considered for styling, starting with the "A" and "B" tractors. This was a unique and new concept in the farm machinery industry, but a technique already being used in the marketing of automobiles and consumer goods. The Henry Dreyfuss design studio in New York was contacted to produce suggestions.

The result appeared in 1938 with both models, and very popular and pleasing this turned out to be. It was well received by customers and lasted for the next 14 years. As a result the smallest and the largest models in the line, the "L" and the "D," were both styled for the following year, and the new small tractor, the "H," appeared with "A" and "B" type styling from its introduction.

A Small Row-Crop Tractor

The demand for a tractor of this size arose from the Allis-Chalmers announcement of their Model B, and some 60,000 "H"s were made over their 9-year production life. There were no significant mechanical changes during this time.

A Model "HN" was added with a single front wheel instead of the standard twin front version, and in 1941-1942, the "HNH" and "HWH" units were added for one year only. Almost all the "HNH"s and all the "HWH"s were shipped to California for use in its bedded crops. They had Model "B" pressed steel rear

1940 Model "HN" No. 29,082 with single front wheel and
hydraulic lift options, at Devizes, England, in September 1981

wheels, 38″ as opposed to the 32″ of the "H" and "HN," with special hubs with 9-stud wheel bolt pattern instead of the 7-stud of the 32″ type. The "HWH" was also available in two types, some having a long front axle casting, H968R, and others a shorter version, H967R.

Some of the last Model "H"s were fitted with electric starting and lighting. Again while the early models had a curtain to control water temperature, the later ones could be fitted with a manually operated radiator shutter like their bigger brothers. Other options were a power takeoff and a simple hydraulic power lift attachment.

Styling for the Largest Row-Crop Tractor

Last of the tractor line to be styled was the "G," but not until 1941. Wartime restrictions were in force by then, and when Deere applied for a price increase to cover the added costs, not just of the styling but also the other updates like the provision of a 6-speed transmission, this request was turned down, since the model name was the same. As a result the newly styled "G" was renamed the "GM" (modernized), and

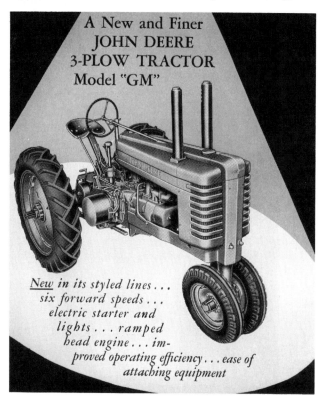

A New and Finer
JOHN DEERE
3-PLOW TRACTOR
Model "GM"

New in its styled lines . . . six forward speeds . . . electric starter and lights . . . ramped head engine . . . improved operating efficiency . . . ease of attaching equipment

The sales literature cover page announcing the Model "GM" in April 1941

remained so called until after the war and the removal of restrictions. The serial number plates showed "Model GM," but the serial numbers remained No. G-13,000 series.

These tractors can be easily distinguished from their smaller counterparts by the side-by-side position of their exhaust muffler and air intake, as on all the unstyled models, as compared with the in-line option adopted for the "A" and "B" series.

They can also be distinguished from the later electric-start "G," as the model again became known on de-restriction, by the retention of a pan-type seat. The later "G" was available with either a single front wheel or a wide front axle, these being of the interchangeable type pedestal later adopted on the electric-start "A" and "B." Unlike them, the electric "G" never had the pressed steel side frames.

MODEL *"HWH" for Bedded Crops*

"HWH" announcement in March 1941. The tractor, a "smash hit" with Western growers originally, also scores with collectors today.

Meantime the standard and orchard tractors did not receive the styling given to the row-crop models, except the Model "D." In fact the "BR" and "BO" were never altered during their lifetimes. The "AR" and "AO" did finally become styled in 1949, and that after other developments had taken place.

The exception to the rule was the "AO," which in November 1936 had appeared in streamline form to avoid snagging the branches in orchards and groves, These models were not the work of Dreyfuss, and only remained so "clothed" until October 1940. When 821 units had been so built, the company reverted to the later center pressed steel radiator cap style adopted

View of the Moline branch house office tower from the LeClaire Hotel, with the Mississippi River bridge and Iowa on the far side, in September 1959. The offices later housed the John Deere export division.

for the "AR." This small styling change coincided with the introduction of a larger-stroke engine, up from 6.5″ to 6.75″.

Wartime Problems

Between the beginning of the war in Europe and December 7, 1941, when America entered the conflict, the rules of the Neutrality Act had to be followed.

This meant that the large numbers of tractors and plows sent to Britain, for example, had to be paid for in cash. These machines were used to plow up old pastures and even golf courses in order to provide food for the nation and the troops defending it. Some of the cablegrams sent to order this equipment were two yards long! It makes interesting reading to study the number of Waterloo tractors sent to Britain during this period.

1938	85
1939	341
1940	490
1941	873
1942	572
1943	637
1944	442

Total: 3,440

After Pearl Harbor all American manufacturers were under the strictest control by the War Production Board and Board of Economic Warfare. Nothing could be exported without a license, and commercial shipments almost ground to a halt. Lend-Lease began in late 1942 and usually covered tractors, but also included some plows, cultivators, grain drills and

Studio photo of the "AOS" streamlined orchard model, pictured in October 1937.

combines. Tractors were shipped in lots of ten plus a kit of spares.

One Lend-Lease contract covered 1584 "A" tractors. Another was for 1000 "H"s, which went to Russia in 1946. As Bob Lovett, a retiree from Deere's export department, described the situation, "It took a long time, like four years, to beat enough swords into enough plowshares, and plows, and tractors, and grain drills, and combines to fill the pipeline. In fact it is true to say that Deere did not catch up with demand until 1950."

No new production tractors appeared during the wartime period, although experiments were continuing on the possibility of fitting a diesel engine to the Model "D" tractor.

With the success of the Ferguson system and its built-in hydraulics and draft control, it was apparent that a new small tractor was needed to compete with this development.

The Small "M" Tractor

By 1944 the preproduction models of the new small tractor were at work in the field, but their manufacture had to await a further development. In 1947 the company decided to build an entirely new factory on the banks of the Mississippi River at Dubuque, Iowa. The author was privileged to visit this new works just after

Experimental Model "M" tractor in 1944, with the 10AW 5′ cut combine, cutting rye. The 3½′ 10A model proved too small for the market, so the 11A 5′ header was tried as an option. The resulting machine was not marketed.

it had opened in the fall of that year, and as the only plowman in the party, to demonstrate the new "M" tractor with 2-furrow mounted "M-2" plow attached.

The new model retained the two cylinders of the Deere tractor family, but in this case followed the pattern set by the "L" series and used a vertical motor. It was offered to the under-100-acre farmer—more than three million of them in the States alone—as his main tractor, and as a backup tractor for the larger farms.

The new Dubuque Tractor Works in 1947

Following the styling of its bigger brothers, it offered a complete new system of speedy, efficient, modern farming with its Touch-O-Matic controls for integral equipment. Self-starting and power takeoff were standard, and belt pulley and lights were available as extras.

Mounted implements included one- and 2-furrow plows, a single-furrow 2-way plow, disk tiller, disk plows, integral toolbars and spring-tooth harrows, one-row cultivators for use in corn, cotton and other similar crops, planters, listers, and a mounted mower for the forage farmer.

Hydraulic Powr-Trol Introduced

In addition to the "M" with its own specialized hydraulic system, the larger members of the John Deere tractor family were now available with remote control

Powr-Trol on styled "D" showing rear hydraulic outlets

hydraulics, allowing a smooth and effortless raising and lowering of drawn equipment, most of which was also available on rubber tires. In addition, mounted equipment could now be raised or lowered in gradual stages with the new hydraulic Powr-Trol, instead of the up or down of the previous system. This added greatly to the overall flexibility of the tractor's operation.

New Plows Introduced

The old faithful Nos. 5 and 6 plows, in production for over 30 years, had finally given way to the new Truss-Frame 55 and 66 models. Similarly the 52 and the No. 4 series had become the 44, and the large No. 7 was superseded by the 77. This was Deere's biggest and

Model 66 5-furrow Truss-Frame plow pulled by 420C Crawler

strongest model, with 23″ vertical and 24½″ fore-and-aft clearance in either 3-, 4-, or 5-bottom sizes. Their standard equipment included new heavy-duty totally enclosed power lift clutch and spring-release hitch.

Disk plows had increased in size, cutting up to 6′ wide and 16″ deep. Two-way plows had Truss-Frame design like their one-way counterparts; the 202 2-bottom was popular for deep work, as it was easily reducible to single furrow operation. Listers, bedders and middlebreakers were available in integral and drawn types to match the four sizes of row-crop tractor.

Model 101 single-furrow 2-way plow behind a styled 4-speed Model "B" in August 1939

1100 Surflex disk tiller behind a 70 diesel standard tractor in November 1954

Tillage Tools Expanded

Disk tillers were popular in many areas and were now offered with planting attachments, allowing a single operation from stubble to seedbed. There were no less than 85 different types of tiller in the line, with the largest being the 970 series cutting up to 15', although the Surflex 8' tiller could be extended to 20' by adding three 4' extensions. Flexible disk tillers were introduced for the first time in 1947. The first hydraulically controlled wheel-carried disk harrows had been announced the previous year.

The "CC" cultivator made in the Van Brunt Works remained throughout the 10-year period from 1937 to 1947, but disk harrows had multiplied. Here again, though, the popular Model "S" had survived. The standard weight "J" series and the heavy-duty "K" series were probably the most popular, other models having more specialized applications.

Killefer Purchased

One significant move that added considerably to the John Deere tillage line was the purchase of the Killefer Manufacturing Corporation of Los Angeles in 1937. Their deep tillage equipment had sold well in California, and added another facet to the expanding variety of implements available to Deere's farmer customers. Included in the range were subsoilers, panbreakers, and ditching and land-leveling equipment, in addition to the heavy-duty disks and cultivators.

The Quik-Tatch System

A multiplicity of mounted implements and trailed harrows rounded out those necessary for working the ground both before and after planting. The ability to attach and detach this mounted equipment with the John Deere Quik-Tatch system gave farmers maximum adaptability for their cultivating chores. Rod weeders, built big and rugged, helped control weeds in summer fallow, and rotary hoes, in up to 6-row sizes, helped the early post-planting row-crop cultivations.

Planters and Drills

In planters, the ubiquitous 999 finally gave way to the first high-speed 290 2-row and 490 4-row tractor

Model 490 planter with fertilizer attachment behind a 6-speed Model "B" fitted with electric lighting and starting as an optional extra, in May 1941

corn planters. Capable of check-planting up to 60 acres a day of all kinds of corn, beans, peas, sorghum, and many other crops in rows spaced 36″ to 42″ apart, they suited the large farmers. Operating at speeds up to 5 mph when check-planting, faster when not, their

The Van Brunt Works at Horicon, Wisconsin, in August 1939

Picture of 28-row 6″ spaced EN grain drill in June 1941. The tractor is a 6-speed Model "A."

high-speed valves, Natural-Drop seed plates, power lift with delayed-action drop, and automatic wire release soon made them favorites.

For precision drilling of beets and beans, the 6-row 66 low-can drill was a similar tool. It had replaced the 18 introduced in 1938. The one- and 2-row potato planters had been refined and given 12-arm picker wheels, and many owners had reported 98 to 100% accuracy at speeds up to 5 mph.

In 1942 the "FF" combined grain and fertilizer drill with a steel box had replaced the original "F" wooden box drill. The "EN" had been added to the "EE" series, and the "PD" plow-press drill in 6-, 8-, and 10-row 6″ spacing was introduced to pack soil over the seed with its press wheels. This promoted even germination and ripening of the subsequent crop.

From 1944 on most of the drill models were supplied on rubber tires which matched their ability to be operated at speeds up to 6 mph. These represented the first low-wheel drills to be offered to farmers. The Model "A" fertilizer distributor and the "H" lime spreader remained in the line, unchanged by the developments taking place elsewhere.

Deere's first automatic wire-tying pickup baler, the 116W, pictured in a July 1945 hayfield

Haymaking Modernized

In the hayfield the "old faithful" No. 5 mower was still the mower of the decade, a large number being exported in addition to those used widely at home. The "M" tractor had its own 51 fully mounted mower, but the Big 4 tractor-drawn mower was still a favorite in its 4½' to 7' size range.

Rakes and hay loaders hardly changed over the decade, but the introduction of the windrow pickup press in 1936 had revolutionized the work in the hayfield. As mentioned in the previous chapter, the addition of rubber tires and a new power plant had given this machine greater flexibility, but it was still the tool for a large farmer or custom operator. Accordingly, for the 1942 season, Deere introduced a farmer's model with power-takeoff drive, as a worthy partner for the larger heavy-duty model.

An Automatic Pickup Baler

A successor to both these machines, the 116W automatic wire-tying pickup baler, was introduced in 1946. It was the world's first automatic wire-tie baler. Again it completely altered the accepted practices in the hayfield. With only the tractor operator required instead of the two to four men on the hand-operated windrow pickup press of the thirties, this released farmhands for collecting bales from the field.

It was driven from the power takeoff normally, although first a Wisconsin engine and later a 4-cylinder John Deere power unit were available as optional extras. Some 3500 were produced the first year, but all were recalled to the Ottumwa Works for a modification to the wire-tying system that next winter. It was found that movement in the bale chamber caused a problem with the wire twisters, so they were mounted in a large single casting to overcome this.

By 1947 this operation had been completed at no small expense, in pursuit of the company's policy of standing behind its products.

Forage Harvesting

Becoming increasingly popular with the livestock farmer was the production of forage from both grass and corn. Ensilage harvesters and blowers had been introduced by the company in 1938 for the corn crop, and these were updated over the next few years. In 1942 a pickup hay chopper was added. The 62 and 64

An early hay chopper behind a 6-speed "A" tractor in September 1943. The steel wheels of wartime on the forager and the old wooden-wheeled wagon make an interesting picture.

combined the hay and corn chopping machines into one unit with two different heads in 1947, and at the same time the No. 2 auger-feed blower replaced the older conveyor-fed type.

The Combine Harvester Line Expands

It was in the combine harvester line that developments were most marked at this time. The small No. 6 was not a great success, and in 1939, and for that year only, the 4'-cut 10, 5'-cut 11, and 6'-cut 12 combines with right-hand-cut headers were produced. These were full-width, straight-through machines with rasp bar cylinder and optional PTO or engine drive, and with bulk tank or sacker. Styled by Dreyfuss and mass-produced, they were ideal for the small farmer, with one problem only.

Most binders were left-hand cut and the new machines had to operate in the opposite direction, as had become standard with combines over the years. Allis-Chalmers had realized this and had introduced their All-Crop machines with left-hand cut. Accordingly from the 1940 season on, the John Deere 10, 11, and 12 combines became the 10A, 11A, and 12A, with left-hand operation. The 10A was too small to be very popular, and although Deere experimented with a unit fitted with the wider 5' head of the 11A and called the 10AW, this machine was not put into production.

An Auger Platform

Also in 1939, a new combine with a 12'-cut auger header, the No. 9, was introduced with styling similar to the smaller machines. This new approach for conveying the crop to the threshing mechanism

Preproduction Model "M" and 10A combine cutting oats in July 1945 on a farm near Moline

pointed the way for more exciting developments to come. The No. 9 was available for either PTO or Hercules engine drive.

1937 had seen the introduction of the 35 hillside combine, a smaller 12′ or 14′ version of the popular

No. 9 12′ cut auger-header combine with styled "D" in charge near Sydney, Nebraska in the 1941 harvest

36, and a replacement for the Caterpillar 34 in the West. It was no surprise, therefore, when in 1940 Deere announced the 33 hillside 10′ cut to take the place of the Caterpillar 38. The styling was again similar to the No. 9, but it was only available with the Hercules IXB5 engine drive.

One further model appeared before wartime restrictions had their inevitable effect. The No. 7A

Caterpillar D7 and No. 36 Hillside combine in mountainous Pacific Northwest country

replaced the No. 7 in 1941, after it had been in production for nine years, but the new model was short lived. The war conditions demanded simplification, so production ceased on the 10A, 11A, 5A, 7A, 33, and 35, and was interrupted on the No. 9.

The Self-Propelled Era; The Famous Model 55

Experimental No. 44 self-propelled combine based on the 12A threshing mechanism in August 1944. This machine was never put into production. Note the LA type wheels used at the rear.

Experiments, however, were continuing. The 10AW has been mentioned, and another which did not come into production was the 44, a self-propelled version of the well known 12A. Much more important were the experiments being made on a larger self-propelled combine based on the No. 9 threshing unit. Known from the start as the 55, these various trial machines were used extensively in the 1944 harvest season. Built both as tankers and baggers, with both auger and canvas headers, the early models had the engine along the left side of the machine as did the original International Harvester self-propelled machines. They also had the similar offset (left side) driving position.

Standard Layout Adopted

The first production combines were available for the 1946 season, by which time Deere's standard format for self-propelled models was established for many years to come. An auger header, with retracting fingers in the auger to prevent wrapping, and with another beater behind the auger again having retracting fingers, gave the smoothest possible flow of material to the cylinder. This was of the rasp-bar type, although a spike-tooth option was available for certain conditions, and this was followed by long, deeply stepped aggressive straw walkers. The chaffer and sieve oscillated in opposite directions, to prevent bridging of short straws and to give a balanced sieve box.

The central driving position, with grain tank or sacking platform behind and the engine mounted across the machine behind that, led to a balanced load on the drive wheels. The variable ground speed was initially hand operated, but the header was raised and lowered hydraulically.

The World's Most Copied Combine

The ubiquitous 55 was the most copied self-propelled combine of all time. A silver painted model was to be seen for many years in the Claas factory in Germany. Their first self-propelled model was announced in 1953 as the SP 55. Similarly in Belgium, Claeys had a 55, and their first self-propelled model, the MZ, had many castings on it with the John Deere casting number still showing. It was effectively a 55 widened from 30″ to 40″ in the separator for European straw conditions. Claeys became Clayson, then New Holland Clayson, and now Ford New Holland. The parentage of their combines is still obvious for all to see.

A later example 55C experimental combine with top mounted tank and engine, now mounted behind, and with horizontal separator housing

To copy is the greatest compliment one can pay. An interesting story was told to the author by W.E. (Bill) Murphy, known to his friends as "Murph," who spent his life traveling the world with John Deere combines. He was in Argentina with a 55 and Tom Carroll of Massey-Harris was staying in the same hotel. Tom was recognized as one of the world's leading

An early example of the experimental 55 self-propelled combine with the sloping type separator housing of the No. 9 tractor-drawn type, 1944

combine designers, having been responsible for the Massey Models 20 and 21.

He asked "Murph" if he could see the new machine working. Although it was not intended to take it out that day due to miserable weather conditions, they went out to the field. The new combine had only traveled about 100 yards when Carroll said, "Put it away, I've seen all I want to see." It was 1949 before Massey introduced in their header augers the retracting fingers which Carroll had wanted to see in operation. The plain augers without retracting fingers were notorious for their wrapping in damp conditions.

A picture of a production 55 working in a July 1946 harvest field. This was the first year of production of this most successful of self-propelled combines. Note the twin drive wheels.

55 Versatility

Only two 55s were exported to Britain, both in 1948, but both came into the author's ownership when first traded in, and one is currently being restored by a friend. The 55 was certainly the leader in its field and as Murph said, had virtually no teething problems from the word go, a tribute to its designers.

It was available in 1947 as a rice combine, the 55R, which could have either oversize tires or crawler tracks, and with steel and rubber feed-rolls to assist in an even feed to the cylinder.

Windrowers

It had been the practice to use the header of some combines, suitably converted, to windrow grain, but in 1935-36 special windrowers were introduced for this purpose. Originally ground or PTO driven by any 2-plow tractor, they were marketed in 8', 12', and 16' sizes.

To pick up the resulting windrows, Deere had introduced in 1940, in addition to the rotary-type pickup attachment for their combines, a belt type with 3-ply belts. Each belt had six rows of light spring fingers which operated with a gentle brushing action, saving more grain or seeds.

Threshers

Although late on the scene as a thresher, the Light-Running John Deere of the early 1930s had a large capacity due to its extra width in the separator. But in 1938 the design was changed from a rack-type separator to straw walkers, and the width in the threshing area was accordingly reduced from 50″ to 42″ in the larger model and from 46″ to 36″ in the smaller. Similarly the cylinder width of the smaller model was reduced by 2″ from 24″ to 22″, but the larger remained at 28″.

Corn Pickers Updated

In 1938 also, the 20 2-row corn picker was replaced with the 21, and two years later the 25 push-type 2-row with the 25A. At the same time a corn picker for the small farm, the 101 one-row, was introduced, capable of operation with the Model "H" tractor.

After the war, the 21 was replaced with the new 200 pull-type corn picker, while the 25A became the new 226 mounted machine.

No. 200 corn picker in October 1949 behind a late type electric-start "G" tractor

Potato and Beet Machines

In the potato fields, Deere had introduced their Level-Bed style elevator diggers in both single- and double-row models in 1939, but it was not until 1944 that a serious attempt was made to produce a complete commercially viable beet harvester. This was the 54A integral beet harvester for the "A" or "AN" tractors, capable of windrowing beets and tops separately.

For use with this machine a beet loader was available, while both one- and 2-row beet lifters could be purchased for use with any John Deere row-crop tractors, and a 2-row pull-type for use with any model or make.

Two- and 4-row bean harvester attachments for John Deere cultivators for cutting and windrowing beans were added to the harvester line, and a similar 2-row unit was available for peanuts.

The Cotton Crop

The early cotton harvesters had given way to the 2-row integral 15, a stripper which represented a large saving in harvesting costs over hand work—$30 per bale or more in 1947.

A 2-row Level-Bed potato digger behind an "AW" tractor in October 1936

No. 15 2-row cotton harvester mounted on an electric-start Model "B," pictured in December 1947

Shelling corn with a No. 7 sheller, belt driven by a "B" in February 1944

Miscellaneous Machines

Having harvested the grain and corn, those farmers who wished to utilize part of their crop for their own animals had requirements for roughage and hammer mills. The Letz line of the thirties and before had been replaced by 10- and 14-inch hammer mills in 1937-38. By 1947 the No. 6 6-inch, the 10A 10-inch, and the 14 high-capacity 14-inch were listed. The 110A and 114A roughage mills were similar basic units with a different feed table, suitable for all types of forage and fodder crops.

Corn Shellers

The corn grower was well supplied with barnyard equipment, and had been throughout the years 1937-1947. The No. 1 hand sheller had become the No. 1-B, and with the cylinder shellers, the No. 4 became the No. 4-B, while the Nos. 5, 9-A, and 10 had become the modernized Nos. 6 and 7.

Elevators and Spreaders

A Bridge-Trussed portable elevator had been added to the models that had been in the line for many years. Manure spreader models had been updated and were now of modern appearance and fitted with rubber tire equipment. The Model "H" 2-wheel and "HH" 4-wheel were ground driven machines with combined wood

An "A" tractor in charge of an "L" ground-driven manure spreader

and steel boxes, the latter replacing the long-lived and very popular Model "E." A power-driven 4-wheel spreader had been in the line since the mid-thirties, and was not remodeled until later.

Loaders

To load the material, a rear-mounted loader was introduced in 1940 for use with the "A," "B," and "G" series tractors. To balance this a weight box was provided to attach to the tractor's front pedestal. This loader was in turn replaced by the front-mounted 25 push-type, with drive taken from the tractor flywheel, thus obtaining "live" controls.

Deere's first manure loader mounted on rear of "A" tractor in August 1939

Wagons and Trailers

Transport facilities had really changed little between 1937 and 1947. The 4-wheel 951 trailer gear had become first the 952 and then the 953, the heavy duty 902 in the 5000 to 8000 lb. class had been replaced with the 963, and an economy model, the 943, had been added to the line. These three models made up the "Big 3" in Deere's modern all-steel rubber-tired farm wagons. Pressed steel wheels had taken the place of the earlier spoke type, but visually there was little other difference.

A ruggedly built freight trailer, the 16 2-wheel, capable of carrying up to 3 tons load, and fitted with a tilting platform, dual wheels, automatic brakes, and a powerful loading winch, all as standard equipment, was an interesting addition to the trailer line. The 20 trailer cart for use with the farmer's car had been upgraded to the 50 utility 2-wheel trailer.

To satisfy the requirements of the increasing number of farmers with this rubber-tired equipment, a John Deere tire pump had been introduced. This could be driven from the PTO shaft, or there were two models available for flywheel drive, for the enclosed or open type.

Killefer

Killefer had added to their line a landleveler, while the recently acquired Lindeman factory offered land shapers in 20' (LS200) and 40' (LS400) sizes.

An electric-start Model "B" does a good job with a Killefer "8ML" land leveler in February 1948

Horse-Drawn Equipment Still in Catalog in 1947

Last but by no means least, and despite the effects that war had on the acceleration in farm mechanization, a full range of horse-drawn equipment was still offered in the 1947 catalogs. This included all types of harrows, planters, drills, mowers, rakes, hay loaders, binders, potato diggers, manure spreaders, and wagons, much of this now rubber tired and updated to match the tractor-drawn equivalent.

The company's product line was in good shape to face the next few years, when its 2-cylinder tractor line would reach the zenith of its capabilities. The "full line" was becoming ever more complete.

1947–1959 end of an era

Rapid expansion after wartime restrictions. A crawler tractor in the line. Deere's first diesel tractor. Increases in horsepower demanded. The climax of the 2-cylinder era. New range of twine and wire-tying balers allows one-man haymaking. The whole combine line modernized. Self-propelled windrowers and cotton pickers. William A. Hewitt assumes command and develops international outlook.

Expansion of Tractor Line

The year 1949 saw the most significant developments in the John Deere tractor line since the introduction of the famous Model "D" some 25 years previously. With the new works in Dubuque fully operational, the company was able to expand the "M" line of tractors to include a tricycle model, much as had occurred 20 years earlier with the then smaller tractor, the "GP."

A studio picture of an "MT" with wide adjustable front axle, October 1949

In addition to the resulting "MT" twin-front-wheel model, both a single-front-wheel version and one with an adjustable wide front axle were announced. More significant in the long term was the introduction of a crawler to the line. The "M" skid unit was built in Dubuque and then shipped to Yakima, Washington, to the recently acquired Lindeman factory. Here it was fitted with crawler tracks and became the "MC." With a crawler tractor in the line the possibilities of developing a range of industrial machines moved that much closer.

Rear view of a styled "AO" orchard tractor, showing the protection for branches, the hydraulic Powr-Trol unit below the seat, and full-fender wheel protection

Two tractor models which had missed out on the Dreyfuss styling exercise, the "AR" and "AO," were remodeled in 1949 with a new look. This new look was adopted at the same time in a new tractor, the most far-reaching development yet—Deere's first diesel.

The Model "R"

Experiments toward this goal had been in the pipeline for ten years, a period inevitably extended by the intervention of the war. The MX experimental tractors finally came close to their production profile in 1947 and were much talked about during the writer's first visit to Waterloo in the fall of that year.

The "R" was introduced to dealers in June 1948, and became available for the 1949 season. It was an immediate success, and with the other tractors in the line gave farmers the choice of 18 models. With the further option of gasoline, all-fuel, and diesel engines, it was possible to match the exacting needs on farms of every size and type.

Hi-Crop
Tractors Introduced

By 1951 the number of models available had increased to 20 with the introduction of the "AH" and "GH" Hi-Crop models. These units were now practical because of the standard use of electric starting, in view of the high position of the engine's fly-wheel. With a 32″ clearance underneath, and 48″ between final-drive housings, they were ideal tractors for use in tall-growing, bushy or bedded crops.

Model "R" No. 15097, photographed in October 1981. This tractor is one of four Model "R"s in the United Kingdom.

Model "A" High-Crop introduced in 1950; shown here driven by Matt Dewatter at Layher Farms in Wood River, Nebraska

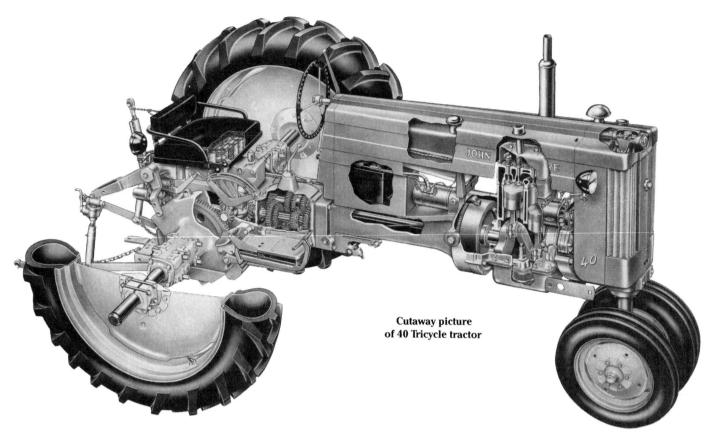

Cutaway picture
of 40 Tricycle tractor

The Numbered Series

The demand from farmers in the early fifties was for increasing speed of operation, which involved higher horsepower from their tractors. Accordingly in 1952, Deere's two most popular models, the "A" and "B," were replaced with the new numbered 60 and 50 series tractors. All the optional versions, in both engine and front-end assemblies, were continued. Extra power was achieved from the same size

engines by using duplex carburetion, hot and cold adjustable manifold, and cyclonic fuel intake.

In addition "live" powershaft and "live" high-pressure hydraulic Powr-Trol were added as extra options, and rear wheel tread adjustment was redesigned to give quick change for different tread widths. Both models could have a downswept exhaust to give maximum clearance, the air intake having been moved to a position behind the new style, easy-to-clean grille, which was similar to that of the "R."

An assembly of numbered series tractors. Left to right, an 80 with factory fitted cab, a 70 diesel row-crop, a 60 orchard with front protection grille and full rear fenders, a 50 in rear, and a 40 row-crop utility.

The following year saw the update of the "M" series to the 40 series, with the same size engine but governed 200 rpm faster at 1850 rpm. The "G"s became the 70 series, and these featured all the options available for the 50 and 60 tractors, including high-compression gasoline or the more usual all-fuel engine. The cast-type frame of the "G" was still used.

The 40 Crawler had a 4-roller track frame instead of 3 as in the "MC," and an additional model was added, the 40 Utility, effectively an orchard or grove version since it was some 8″ lower than the standard model.

The 3-Point Linkage Era

Bob Hanson at the wheel of a 60 tractor with integral 813 3-furrow plow in 1964

The 3-point hitch on these smaller tractors took implements of any make and retained a swinging drawbar for trailed implements. It was fitted as standard with Deere's exclusive Load-and-Depth Control. The hood and grille were easily removed for cleaning and servicing.

No less than 37 items of mounted equipment were available including plows, tillers, disk and other harrows, the No. 1 tool carrier, bedders, listers, planters and drills, cultivators, a rotary hoe, a mower, loader, forklift, snowplow and rear scoop—a dazzling array of implements to satisfy every farm need.

In the early fifties, first attempts to attach mounted equipment to the larger tractors featured the introduction of the 2000 series tool carriers, followed by the 2100. Greater flexibility was achieved with the announcement of the 800 hitch in 1953, which made

pick-up-and-go farming quicker, easier, and more practical than ever before. More than two dozen implements could be adapted for use with this hitch.

LP-Gas Engines Added; the "D" Line Ends

1953 saw the addition of a 5-roller 40 Crawler to the tractor options, and the further choice of engine to burn LP gas for all the 60 and 70 models. It also saw the final demise of the Model "D"—a few "streeter" models being the last made. These were so called because they were assembled in the "street" between the truck shop and mill room in the Waterloo Works. Their serial numbers are 191,579 to 191,670, the latter being shipped on March 18, 1954.

In the case of the standard models, the 60 Standard was initially similar to the styled "AR" with a low seating position, but in 1954 it was redesigned to be similar to both the row-crop models and the 70 Standard with their high platform and seating.

The "R" stayed in production until 1955, when it was replaced by the new 80 diesel. With its engine bore increased from 5.75″ to 6.125″ and the governed engine speed increased from 1000 to 1125 rpm, its power on the belt went up from 51 to 67.6 hp.

A standard 70 LP-gas tractor plowing with a 66H plow on rubber tires in August 1955

Wide-tread 70 diesel tractor with 66 4-furrow plow, seen in May 1955

A Diesel Row-Crop Tractor

Just prior to the introduction of the 80, the 70 was offered with a diesel engine in both its row-crop and standard forms. This represented Deere's first diesel row-crop tractor. An example of this version of the 70, when tested at Nebraska, wrested from the "R" the title of most fuel-economical tractor tested there.

Both these new diesel tractors had a V-4 donkey starting engine in place of the 2-cylinder horizontally opposed unit of the "R." The introduction in 1954 of factory-engineered full power steering on the 50, 60, and 70 models, followed the next year on the 80 when it was announced, represented another enormous advance for the operator's comfort and well-being.

Three new versions of the 40 series appeared for the 1956 season. A 2-row utility low-built tractor was added to straddle two rows in planting and cultivating. A 40 Hi-Crop for use in tall, bushy, or bedded crops, similar to the 60 and 70 Hi-Crops, was announced, together with a 40 Special for use in cotton, sweet potatoes, corn, cane, and other high-bedded crops.

William A. Hewitt, president of Deere & Company 1955-1964 and chairman 1964-1982

Enter the "20 Series"

The demand for increased power continued unabated; therefore, in 1957 the whole line—now consisting of 30 basic models in six power sizes—was increased in power by approximately 20%, and renamed the "20 series."

Distinguishable by the yellow panel along each side of the hood, they were a good looking family of tractors. With the upgrading of the 40 series to 420, a need was felt for a lower-powered machine at the bottom end of the line. So what was essentially the 40 was continued in production and called the 320. This model was available in both standard form and the lower-profile utility.

In addition to the extra power, independent power takeoff—foot controlled in the case of the larger tractors—a Float-Ride seat, and a universal 3-point hitch with Powr-Trol control was offered. This 3-point hitch was also available on the crawler models.

A 5-speed transmission was added to the 420 series, both wheel and crawler, giving extra flexibility. When so equipped, the tractors could be purchased with a continuous-running independent PTO, operated by a 2-position foot clutch.

An October 1956 picture of a 720 diesel standard tractor, chisel plowing with a 650 tool carrier in stubble

The Largest 2-Cylinder Tractor

The initial exception to the overall power increase of the "20 series" was the 820. To begin with, it retained the 67.6-hp engine of the 80, but this was soon remedied with the introduction of the most powerful 2-cylinder engine made by Deere. Tested at Nebraska in October 1957, it developed 75.6 hp and again proved to be more economical than its predecessor—0.388 pounds of fuel per hp-hour against 0.394.

One year earlier, however, the 720 diesel had scooped the honors by producing a figure of 0.383 (or 17.97 hp-hours per gallon) and this record stood until broken by the John Deere 1650 some 27 years later!

Model 320 standard tractor cultivating corn with an integral 4100 cultivator in August 1956

A lineup of "30 series" tractors at the Waterloo Sesquicentennial show in July 1987. Front row—330 standard and 430 tricycle. Rear row—530 row-crop, 630 standard with LP-gas engine, 730 diesel row-crop with Roll-O-Matic front wheels, and 830.

The "30 Series" Finale

With all these options available, it was difficult to imagine a farmer's requirement that could not be satisfied, but after only two years, the tractor line was again altered to become the "30 series." Although no alterations were made to the engine this time, it was operator comfort and a more modern styling which endeared the new tractors to their owners, and to collectors ever since.

A sloping automobile-style steering wheel, instru-ments contained in a sloping panel for easier viewing by the driver, and fenders for the row-crop 530, 630, and 730 with headlights incorporated—a preview of styles to come in the sixties—all gave this last 2-cylinder series a modern and very acceptable appearance.

The saying "If a product looks right it usually is right" applied to the "30 series" tractors in particular. One recalls the ease with which the 830 pulled a 6-furrow 16″ 777H plow some 8″ deep at the big Marshalltown, Iowa, demonstration in September 1959. The climax of the 2-cylinder era had arrived.

Model 435 diesel tractor No. 436,683 at Waterloo, Iowa

A wide front-axle version of the 730 diesel row-crop tractor plowing with a single-furrow 825 2-way plow

A Small Diesel

Only one other agricultural model appeared before the demise of this great tractor tradition, with the introduction of Deere's first small diesel tractor. The standard 430 was fitted with a GM 2-cylinder 2-cycle diesel engine and became the 435. It proved to be the first tractor tested at Nebraska after the adoption of 540- and 1000-rpm ASAE-SAE standards for the power takeoff of tractors.

The development of the industrial line of both wheel and crawler tractors, although occurring at about this time, will be left to the second volume of this product history.

More New Plows Announced

Truss-Frame plows had been introduced in the early forties. An attachment to place a band of fertilizer on the bottom of each furrow was offered for the 44 plow.

The "H" series was added to the line, with a remote hydraulic cylinder used for raising and lowering the plow in work. Safety trip standards had been introduced on integral plows in 1950, a feature made more readily available with the Truss-Frame design.

In 1953 the company had announced its high-speed plow bottoms, enabling much faster plowing speeds to be achieved than were previously possible. These

were extended first to the 55, which thus became the 555, and soon afterward the 44, 66, and 77 were similarly treated and became the 444, 666, and 777, respectively. At the same time, twin tapered roller bearings for the wheels of this series were adopted.

The whole range of plows, from the one-furrow for the smallest tractors to the largest 6-furrow, were all made to the basic Truss-Frame design by the end of the fifties. This applied to all the 2-way models, including the enormous DT3 heavy-duty 3-furrow designed for use with crawler tractors in irrigated land. With its 18″ bottoms it weighed no less than 4600 lbs.

An 830 makes easy work of a 6-furrow 777H plow raised and lowered from the tractor's hydraulics

Other Tillage Equipment Updated

Disk plows from one- to 6-furrow had rugged overhead frames to give maximum clearance, and could be purchased in both drawn and integral form. Disk tillers were preferred in many wheat-growing areas. The largest of these, the 1000 series of Surflex tillers announced in 1949, could cut up to 20′ in width.

Surflex disk tiller with seeder and packers behind a 720 diesel tractor in May 1957

A wide-front-axle 70 row-crop tractor with "RW" wheel-mounted disk harrows

These were followed by the 2200 series in 1954, and the 2200A in 1956 for deep work with their 26″ disks spaced 10″ apart.

Subsoilers and panbreakers, the heavy-duty ones from the Killefer Works, continued in the line with occasional additions as the need arose. The 23 toolbar became known all over the world and was later made in South Africa.

The popular CC spring-tine trailed cultivator was modernized and became the CC-A in 1958. The C-4 and C-7 3-point-hitch units had succeeded the No. 8 previously in 1955. Rodweeders continued to be popular for summer-fallow work; the 500 series, introduced in 1959, was claimed to be clog free.

Disk harrow models proliferated, from some ten types in the 1947 catalogs to 22 by 1959, and all of these were in many sizes. The first wheel-carried disk harrows, the Model H offset, had been announced in 1946 and these had become the norm by 1959. The Killefer Works was responsible for all the heavy-duty equipment produced, and also made certain specialized items like their land-leveler.

Spring-tooth and spike-tooth harrows became available in ever-wider sizes with the use of squadron hitches, and these could be adapted for pulling multiple units of other tillage and planting equipment. A roller-harrow unit was an innovation in 1959.

New Planters

With the increased power at their command, farmers were requesting larger planters. The 290 and 490 2- and 4-row models were superseded by the 4-row 494 and the 6-row 694 for operating at speeds up to 5 mph. Extras for these new units included dry fertilizer, preemergence herbicide, and insecticide attachments. Farming was rapidly becoming more scientific, and maximum yields increasingly important.

Flexi-Planters mounted on an integral toolbar with 620 tractor in April 1956

For corn and beans the new 246 and 446 planters were ideal, while for cotton and corn the 247 and 447 were supplied. All these were drawn types. The integral units for cotton and corn were the 484 and 684 models.

Two 494 planters with an 8-row hitch behind a 730 diesel tractor in 1959

Drills

Drills had adopted the low-wheel rubber-tired equipment since 1944 and were little changed over the period under review. The popular Model "B," which had replaced the "EE," remained in production

throughout, as did the "PD" press drill. The combined grain and fertilizer model "FB" of 1951 became the "FB-A" in 1955 and the "FB-B" with larger-capacity boxes in 1959. The "RB" was replaced with the "DR" Double-Run model, while the "LL" press drill became the "LL-A" with a new look and larger-capacity hopper. Added to the line was the "LZ" lister grain drill in 1951, and this became the "LZ-A" in 1956 and "LZ-B" in 1959. Two other specialist models, the "RL" Rangeland, for putting new life into grazing land, and the "GL" Grassland, for direct drilling into established sod, extended the drill line. The latter was featured on the cover of the 1955 Better Farming catalog and became the "GL-A" in 1956.

The Model "H" lime spreader had been in the line for a long time, but was finally replaced for the 1955 season with the "LF" Propel-R-Feed, available in 8', 10', and 12' sizes. This spreader was an extremely accurate distributor of fertilizer, and could also be used for small seeds when a different bar-type agitator was used. It was the only John Deere-designed machine built for a time in England. A mounted version, the "MLF," was added in 1958, as also was the "LD" liquid fertilizer distributor.

An 820 comfortably manages two 24-row grain drills in this 1957 photograph

Potato Machinery

The one- and 2-row potato planters with 12-arm picker wheel remained in the line and were updated in 1954, a 4-row model being added. The Level-Bed diggers for harvesting the crop were in one- and 2-row form; the 30, a Double Level-Bed digger with two 26″ elevators side by side, solved the problem of trashy conditions.

A 4-row potato planter pictured in July 1954 on rubber tires

Sprayers Introduced

The year 1959 saw the introduction of drawn and mounted sprayers in both tank and drum types. The Models 10, 20, and 30 were for one, two, and three

55-gallon drums, with the 30 supplied alternatively with a 200-gallon aluminum or steel tank. All models could have 6- or 8-row booms, which could be folded for transport. A hand gun was offered for spraying at close quarters.

Haymaking Equipment Modernized

Mowers had not changed for 20 years or so except for the addition of rubber tire equipment. It was not surprising, therefore, that with the increased speeds now popular and commonplace with farmers, these long serving machines should be updated. First the 51 rear-mounted mower for the "M" tractors was replaced by the center-mounted 20, available for the new 40 tractors as well. This unit was in turn replaced by the 20-A in 1957. The next year the long serving and excellent No. 5 was replaced by the No. 8, with additional features and up to 9′ cut. The No. 9 was a fully mounted version of the same.

After the introduction in 1955 of a mower-conditioner, which cut hay-curing time in half, a special PTO drive attachment was available to drive through the No. 8 mower to this machine. This gave a once over operation. The same mower was also offered with remote hydraulic-cylinder lift.

Model 30 sprayer mounted on a 530 tractor spraying alfalfa in July 1958

A mower and conditioner combination behind a 520 gas tractor. The No. 8 semi-integral mower had a PTO option for further trailed machines.

Rake Models Match Modern Needs

Rakes were improved by mounting them on rubber tires (the 594 series) or making them semimounted (the 851), and curved teeth became a John Deere trademark. A new PTO-driven rake, the 350, was announced for the 1957 season. It could be either fully mounted or semi-integral as required.

In 1959 two new rakes capable of operating at up to 8 mph were the 894 and 896. The latter featured the widest raking width on the market. It was capable of following the new 9'-cut mowers. Another new model, the 858, was introduced in the same year for light, short, viny crops.

A 430 row-crop utility tractor makes light work with a 594 side-delivery rake

In a 1957 hayfield, a 620 pulls a 14T pickup baler fitted with an ejector, making a one-man outfit

A Baler Revolution
One-Man Haymaking

The 116W automatic wire-tying pickup baler making 16″ × 18″ bales was joined by its smaller 114W version for those who preferred 14″ × 18″ bales. The big advance came in 1955 with the announcement of the 14T twine-tying baler, which set the pattern for all subsequent John Deere square balers. With its floating auger feed and—for the times—wide pickup, it proved a winner. Normally PTO driven, it could be supplied with a 15-hp auxiliary engine—a 2-cylinder Wisconsin THD gasoline unit.

With the advent of this new baler, 1957 also saw the first bale ejector, introducing the possibility of one-man haymaking. A barn bale conveyor with automatic random stacking of the bales completed this whole operation the following year.

The earlier 116W balers had been available with a Wisconsin V-4 engine in place of the PTO drive. This option was replaced with a John Deere 4-cylinder in-line power unit—a shape of things to come—the 92 series with 29.3 hp.

The 214 Series Balers

These 116W balers, though well liked, were of limited output, and in 1958 were replaced by the new 214T and 214W models. These were heavy-duty versions of the 14T, with unique pressure plates in the bale chamber. The tidiest bales of any on the market were produced due to the four-sided pressure applied.

Both the twine and wire tying models were offered as standard with PTO drive, but reverted to the Wisconsin V-4 air-cooled engine as an option. Both

By comparison with the 14T, the bales from the 530 and 214T baler require collection

80

of these balers were the first John Deere machines to be built in the newly acquired Lanz factories in Germany—but more on that later.

The 14 light power stationary baler was still in the 1948 catalog, as were the A306 standard and A306G heavy duty combination raker-bar cylinder loaders for dry hay and green silage, respectively. Also listed was the 204 double-cylinder green-crop loader.

The Forage Crop

While the 62 forage harvester continued in production until 1954, the No. 2 forage blower was re-

A dusty 70 LP-gas tractor with a single front wheel is in charge of a No. 8 forage harvester in May 1955

placed in 1951 by the bigger capacity 50, with its wider hopper and heavy duty fan.

For the 1955 season the No. 8 harvester was introduced, for the first time offering the choice of three harvesting attachments (headers), plus a windrow pickup and a mower bar. Still of the flywheel type, this new model represented the need to match the increasing tractor power becoming available to the grassland farmer.

A Rotary Forage Harvester

The following year saw another forage innovation with the announcement of the first rotary forage harvester, the 10 chopper. Basically consisting of a transverse rotor with four rows of eight curved, free-swinging knives, it cut a 5' swath and then augered the material to a blower for delivery to a following trailer or truck alongside. In addition to forage, the machine could be used to cut brush and cotton, corn and other stalks, or to collect beet tops. It represented a truly versatile addition to the forage family.

1958 saw several developments. The 10 chopper was replaced by the 15. With the addition of knives to its fan blades, it became the first double-chop forage harvester. It could cover all the operations of the 10, including picking up windrowed crops.

A 620 tractor and 10 rotary chopper with a feed wagon in tow make a clean job of harvesting a tall crop

The Chuck Wagon Arrives

An updated version of the outfit on the previous page. A 630 has replaced the 620, the later 15 rotary chopper superseded the 10, and the 110 Chuck Wagon represents the latest in green crop haulage in 1959.

Two new methods of transporting the forage crop were announced. The 110 Chuck Wagon meant the crop could be unloaded mechanically into the new 55 belt-conveyor blower with its 11½'-long, 30"-wide hopper. In addition, a forage box attachment was listed for the large capacity Model N PTO-driven manure spreader, thus making it a dual-purpose machine.

By the end of the decade it was decided that two forage harvesters would more correctly meet the requirements of the market. As a result the No. 8

was superseded by the No. 6 for the average farmer, and the 12 for the larger acreage or custom operator. Both models were still offered with the three header options. In the case of the No. 6, the mower unit could be either 4' or 5' cut, while the 12 had a 6' or 7' option.

The row-crop header for the No. 6 was listed in two forms, the standard 6 or a lighter 6A low-cost unit. All versions could be operated by a manual lift or by remote hydraulic cylinder, and all could be had with dual-wheel equipment for better flotation.

While the No. 6 was a PTO-driven machine, the 12 could be either PTO driven or supplied with a John Deere 65-hp EA217G power unit. Both machines could have lifters for their mower bars for badly laid crops.

Harvesting Developments

As a link between the harvesting of crops and the mere tidying up of various residues about the farm, the introduction for the 1956 season of the Gyramor rotary cutter represented the beginning of a long line of these popular and robust tools.

In 1958 the new 207 rotary cutter with 5½' cut (7' with gatherers) and a 15" deep body, either mounted or drawn type, replaced the original design. The following year the 127, a smaller 5'-cut 8"-deep light-

The 5-roller version of the 420 crawler tackles some rough country with the original Gyramor version of the subsequently very popular rotary cutter

weight machine, was added to the line, extending the applications and usefulness.

Exit the Binder and Thresher Era

The John Deere tractor binder, both in its grain and corn forms, finally disappeared from the company's sales lists in 1954, as the two thresher models had in 1951. An era involving much hard labor for the farmer, his staff, and his family had finally come to an end.

The end of one era and the beginning of another. A large load of threshers, with 12A and 17 combines, is ready to depart down the Mississippi in 1940.

The Longest Lived Combine

The machine which had been marketed for longer than any other, apart from the Model "D" tractor, was the renowned No. 36 combine. Originally introduced in 1919 by Holt as the first-ever all-steel combine, it was subsequently upgraded in 1923 to the Model 30, and took its final form as the Holt 36 in 1927. At that time it was built by Caterpillar's combine division, the Western Harvester Company of Stockton, California, the Holt home town.

With the adoption of the Caterpillar name in 1931, in place of Holt, the Model 36 finally became a John Deere machine in 1936. With the acquisition of this historic combine line came one of its best known employees, W.E. (Bill) Murphy, better known to his friends as Murph. As already mentioned in connection with the introduction of the new 55 self-propelled, it was the writer's good fortune to meet up with him in Moline in 1959. The resulting exchange of many a good combine story was the inevitable outcome.

The fitting of optional rubber tires on the 36, a change of make of its power unit, and its final appearance in the all-green-and-yellow livery of the rest of the line represented the only changes in a proven design—a truly remarkable record. With the anticipated introduction of the self-propelled hillside combine, the 55H, it was in 1953 that this long-lived model disappeared from the sales manuals.

Two long-lasting machines join forces. The 36 combine in its last form on rubber tires and painted green and yellow is pulled by one of the latest "D"s fitted with electric starting and lights. The crop has been windrowed to prevent storm damage and the combine is fitted with a straw spreader to distribute the crop residue; August 1947.

The very popular 12A 6′ cut combine with bulk grain tank and auxiliary LUC motor with an electric "A" in charge, July 1948

Modern Tractor-Drawn Models

The demand for a trailed version of the highly successful 55 self-propelled, particularly from farmers who windrowed their grain, flax, and other small seed crops, saw the introduction in 1949 of the 65 12′ cut. It was really a No. 9 with a 55 threshing body.

Replacing the 12A in 1952, the 25 combine offered the choice of a 6′ or 7′ cut, PTO or a 4-cylinder John Deere engine drive. Other modern features included quick-change cylinder speed control, open-bar grate, all-steel straw rack, and a windrow spreader when used with the belt pickup. It retained the canvas feed, from cutter bar to drum, of the earlier models.

It was destined to last in production for only four seasons. Its replacement, the 30, was introduced for the 1956 season. It had an auger with retracting fingers, followed by a raddle feed to the drum as in the larger machines. It was 7′ cut only, but retained the straight through form of its very successful predecessors, and remained in the line for a further five years.

A Smaller Self-Propelled

As had been the case with tractors, so with combines. The overwhelming success of the 55 brought immediate requests for a smaller version of this machine. The result was the introduction of the equally popular 45 in 1954, with either 8′ or 10′ cut. This model was to have eternal fame as the first combine to be fitted (in 1955) with a 2-row corn header, at one stroke revolutionizing the harvesting of America's most important crop. For the 1954 harvest, therefore, the company's combine line was streamlined to the 25 and 65 trailed models, and the 45 and 55 self-propelleds.

Deere's second self-propelled model, the 45, picking up a windrowed wheat crop with its draper pickup attachment

A Larger Self-Propelled

The other major achievement of this decade had to await the 1958 harvest when the large 95 self-propelled was added. The option of 14′, 16′, or 18′ cutter bars, with an 80-hp John Deere 6-cylinder engine and all the advanced features already proven on the smaller models, ensured this new combine's immediate acceptance by the larger farmers worldwide.

It was my privilege to import the only example of this model into the U.K. in the winter of 1959, and this

The width of the largest combine in the line in 1957, the new 95, can be appreciated from this August picture

with a 12′ cutter bar, which the Moline export department did not think possible. By adding a 55 12′ reel pack and knife to the 95's 12′ pickup platform, the impossible was achieved for British crop conditions, and has been a topic of comment in certain quarters ever since!

Simultaneously with the announcement of the standard 95 combine, both a hillside version, the 95H, and models for the rice fields, the 95R and 95RC, were added to the line.

Edible Bean Combines

From the 1956 season the 45 and 55 combines were available with spike-tooth cylinders and pickup platforms, and these versions were then designated edible-bean combines, and in 1959 the new 95 was so treated as well.

With the introduction of full power steering as standard on the 55s, the rice machines were offered as the 55R, with deep-lugged 18″ × 26″ ricefield tires, or the 55RC, with crawler tracks. All the self-propelled combines, and the tractor-drawn 65, could have either straw spreaders or the new straw choppers as optional equipment.

In 1958 the 45 combine had the optional rotary pick-up preferred in beans and a straw chopper

A New Line of Combines

As the fifties drew to a close, a new range of combines, the Hi-Lo series, was introduced. Retaining the same model numbers but with several new features and more modern styling, they will be described in Volume 2. A smaller 8′ cut self-propelled, the 40, was added to the line for the small corn and grain farmer, and was first seen at the Marshalltown demonstration.

A baby self-propelled, the 40 8′ cut, joined the list of combines available for the 1959 season. In general layout and options available, like the straw-spreader on the model shown, it matched its larger companions.

Windrowers

With the completion of a full range of combines, the only other grain harvesting machines to consider were windrowers, for those areas which required this method of dealing with the grain crop.

From the 1930s these units had been available with up to 16′ cut, and had been driven from the tractor's power takeoff. In 1957 a self-propelled windrower was added to the line. With a 14′ cut and an economical 23-hp Wisconsin V-4 gas motor, it represented a simple 3-wheel machine capable of cutting 75 acres a day.

An experimental self-propelled windrower with a 4-cylinder Dubuque engine and 3-wheel configuration is shown in September 1955

The Corn Harvest

At the same time that grain binders had been phased out, so too were their corn cousins. Corn pickers had been established long before, but

A portable batch crop dryer, the 458, with a 620 tractor attached in April 1958

the introduction of the combine corn head finally sealed the binder's fate.

An addition to the line for the 1959 season had been the 458 crop dryer. Driven from the PTO or electrically, they were 400-bushel batch-type units. A further portable crop dryer, the 88, was added for the 1960 season.

A Corn Snapper Introduced

The 226 and 200 2-row corn pickers continued in production into the fifties, and the 101 was adapted for use with the new "MT" tractors. In 1951 a corn snapper was announced, the 100. Most parts were interchangeable with the 101, except the husking bed, which was replaced with an auger.

The 100 one-row corn snapper was similar to the semi-integral 101 corn picker but without its de-husking rolls

New Corn Pickers

1954 saw the modernization of the 226 to the 227. It was new in design from stem to stern. Longer and more gently sloping gatherers, new snapping roll adjustment from the tractor, new shields to reduce plugging, new husking beds with four rubber rollers, and wider, deeper dimensions throughout the machine ensured much greater "self-propelled" harvesting output. At the same time a one-row mounted

The fully mounted 226 corn picker gave self-propelled flexibility to the corn harvest

version, the 127, was announced, thus increasing to five the models listed.

Finally on the corn front, the 50 corn sheller attachment for the 227 picker allowed field shelling of corn, or ear harvesting with the same machine for the first time. The 227 was itself updated at the same time, and Quik-Lube Multi-Luber lubrication could be factory installed if one wished.

King Cotton

The 15 2-row mounted cotton stripper, developed from the first 2-row stripper of 1945, remained the base machine for ten years, but 1950 saw the addition of the No. 8 2-row self-propelled cotton picker—another Deere first—picking cotton lint from both sides of each row. This machine revolutionized cotton harvesting, worked well in tall or short crops, and did not strip off the unopened green bolls. It was a true one-man machine.

In 1955 a one-row mounted picker, the No. 1, was announced for mounting on the 50, 60, 70, and late model "A" tractors. Again a one-man machine, it could do the work of 40 hand pickers. The following year the old faithful 15 was replaced with the 16, with new ease of attaching, and easily adaptable for rows from 36″ to 42″.

Both one- and 2-row pickers were replaced in 1958 by the 22 one-row high-drum mounted picker and the 99 series 2-row high- and low-drum self-propelled. Both machines had been tested extensively during the previous harvest, and represented a significant advance on the earlier designs with their Air-Trol picking. It gave whiter, cleaner cotton by keeping fine trash out of the collecting basket, with an air-intake away from the rows of cotton.

The final cotton harvesting development in the fifties was the 11 one-row forward-mounted picker for the 430, 530, or 630 tractors, designed for use on farms with as little as 40 acres of cotton.

In the cotton fields,
the No. 8 2-row picker gave similar
self-propelled advantages

An "AN" tractor fitted with a 200A beet harvester leaves a windrow of beet tops clear of the machine wheels, while it loads the beets into an attendant truck in Colorado in November 1952

Beet Harvesting

The 54A beet harvester of 1944, with its associated No. 6 beet loader, was not as successful as had been hoped, and was dropped from the line in 1948. The 2-row 33M mounted and the 33T tractor-drawn lifters and the one-row 43 model to fit any John Deere tractor filled the need until 1953, when the new 200 2-row harvester was produced.

The following year it was updated to become the 200-A, and the 100 one-row joined it in the sales catalogs. Their speed and efficiency were largely due to the ground-driven steel-tine lifter wheels, which squeezed the beets out of the ground. The new 210 2-row beet topper joined the line in 1956.

Miscellaneous Equipment— Manure Spreaders and Loaders

The early fifties saw a hydraulic manure loader, the 30, announced for the "M" tractor, followed shortly by the 40 and 50 loaders for tractor models up to the 70. 1952 saw the introduction of the first high-speed ground-driven 70-bushel manure spreader, the Model

"L," and shortly afterward, its 4-wheel equivalent, the Model "M." This latter model was offered for either tractor or horse haulage. Two years later the model "N" PTO-driven spreader, with a 120-bushel capacity, joined the line. Its feed conveyor was ground driven, giving an even, uniform spread.

A low cost "live" hydraulic front loader, the 45, was announced in 1955 with Quik-Tatch 5-minute fitting

A 620 with wide front axle, fitted with a 45W front loader, makes short work of filling an "R" manure spreader in May 1957

to any John Deere tractor. The same year saw the introduction of the medium-size 95-bushel Model "R" ground-driven spreader, followed in 1958 by the PTO-driven Model "W," giving a choice of four modern models to cover all requirements. Already mentioned was the forage box attachment for the Model "N," making it a dual-purpose machine.

For farmers with wide-front-axle tractors, the 45-W was added to the choice of loaders in 1959, as the popularity of this type of tractor increased. The smallest and largest spreaders were uprated in capacity that year to 76 and 134 bushels, respectively.

Mills

The two larger 10″ and 14″ hammer and roughage mills were made available early in the 1950s, with PTO as well as belt pulley drive. In addition, a sturdy 2-wheel transport truck allowed movement of the mills from farm to farm.

A 40T with wide front axle powers a new Model 71 corn sheller in April 1955

Other major machines under the barnyard and maintenance heading were the various corn shellers. The Nos. 4-B, 6, and 7 continued in production until 1956, when the No. 7 was replaced with the 71 large capacity sheller, while the No. 4-B became the 43 PTO-driven portable sheller. This new model was supplied with a rubber-tired transporter.

Snow Plows

One other category of implements which became popular with the increasing application of hydraulics was snow plows. For John Deere general-purpose tractors in the early fifties, the 5′ M-60, 6′ MT-72, and 7½′ ABG-90 were front-end blades. These could be used, in addition to snow clearing, for cleaning barn floors, piling manure, leveling aggregates, and doing light grading work. They had Quik-Tatch frames for ease of fitting.

With the introduction of the numbered series tractors, the prefix letters were dropped from the model identification, and the three blades became the 60, 72, and 90, but the model for the 40 Utility tractor was the 5′ 160, and the "M" kept its M-60 unit.

Scoops and Blades

Two other useful tools which appeared in the 1959 catalogs were both rear-mounted. The 20 scoop was suitable for repairing levees, building trench silos, and cleaning ditches and feedlots, and the 80 blade could be angled either way, tilted vertically for ditching, or reversed for backfilling.

Horse-Drawn Equipment

The last mention of horse-drawn equipment appeared in the 1954 full-line catalogs, but the range was still impressively long. From harrows of various kinds it extended to planters, drills, rakes, manure spreaders, and wagons.

By 1951 horse-drawn equipment like this 4-wheel Model "M" manure spreader had a distinctly modern look

A New Horizon

Unbeknown to the great American farming fraternity and their overseas brothers, the 2-cylinder tractor era was drawing to a close. If one had read the signs more closely, the availability of John Deere multi-cylinder engines for combines, balers, and the like was signaling the impending change.

The fifties were indeed the greatest time of change worldwide for the company. From 1951 to 1956 export sales had dropped by 45% from the all-time peak of $24 million achieved in 1948-49. This represented a material figure in the company's income. It became apparent that to obtain a reasonable portion of Europe's vast farm machinery trade, it was necessary to manufacture there. In 1951 it was decided to build a factory in Scotland, but a change of the U.K. government and the consequent refusal to grant priority steel supplies in the difficult postwar years resulted in termination of the project.

On November 1, 1955, the old Export Department was closed, and replaced by John Deere C.A., a Venezuelan corporation. This was the result of U.S. tax laws relating to overseas sales. With the arrival of William A. Hewitt at the head of the company after the death of Charles Wiman in 1955, the export market took on increased importance, since Hewitt had a very international outlook.

The Lanz Purchase

The immediate result was the purchase, the following year, after lengthy negotiations, of Heinrich Lanz in Germany. This gave the company three factories in Germany and Spain—the tractor works in Mannheim, the combine works in Zweibrucken, and Lanz Iberica's works in Madrid. It also gave Deere a permanent base in Europe, which proved to be a great step forward. The integration of this foreign company into the Deere organization will be described in Volume 2.

Developments at Home

At the time of these moves overseas, the company had decided to stop exhibiting at state fairs, as had been the practice in the past, and to hold its own exclusive demonstrations. The last and largest of these was held at Marshalltown, Iowa, in 1959, and the writer was privileged to be there.

The announcement of a large 4-wheel-drive tractor at this show—fitted with a General Motors 6-cylinder engine and with a strange model number referred to as an "eighty-ten" and not an "eight thousand and ten"—must have started dealers and their farmer customers wondering…. The arrival of the answers to these questions must also await Volume 2.

A fully mounted 8-furrow plow must augur some big development for the future…

PART II

PRODUCT REVIEW

Tractor Review

The Beginning—1892

Although John Deere dealers' interest in tractors began to appear about 1910 with the addition of the Gas Traction Company's Big Four 30 in some of the branch house catalogs, it is the Waterloo factory's lineage, dating back to 1892, which must come first historically.

Froelich's decision to form the Waterloo Gasoline Traction Engine Company, along with others interested in building and marketing his successful traction-unit alternative to the steam engines of his day, proved to be ahead of its time. It was not until 1911 that the farmers' interest in tractors reached the point where the renamed Waterloo Gasoline Engine Co. decided once again to test the market.

Waterloo Developments—1912

It first introduced its large Standard 25-hp tractor in 1912, followed by a crawler-tracked version the next year. The year 1913 saw several developments. The introduction of the Light 15-hp tractor was the most important. Its similarity to the Waterloo Foundry's Big Chief has already been mentioned.

The Model "C" all-wheel-drive and the larger "H" 25-hp tractors were advertised that year, though the numbers produced are not known. Production records of the "L" and "LA" tractors are in the company's archives, and details of these are given in the appendices.

The following year, 1914, saw the change from a horizontally opposed 2-cylinder engine to the twin-cylinder type used for the next 46 years. In this case, however, the motor was placed with its head facing the rear of the tractor.

A

B

C

92

A The original Froelich tractor of 1892, shown at Froelich, Iowa, before dispatch to South Dakota for its first threshing season

B Waterloo Boy Model "Sure Grip, Never Slip" 4-cylinder 1913 tractor

C Waterloo Boy Model "TP" 4-cylinder 1913 Standard wheel tractor

D Waterloo Foundry Company's "Big Chief" 8-15 tractor

E Waterloo Boy One-Man "L" tractor with 2-cylinder horizontally opposed 6″ × 6″ 15-hp motor

F Model "L" disking at a demonstration in June 1914 organized by the Arbuckle-Ryan Company of Toledo, Ohio, and attended by more than 200 dealers and their farmer customers

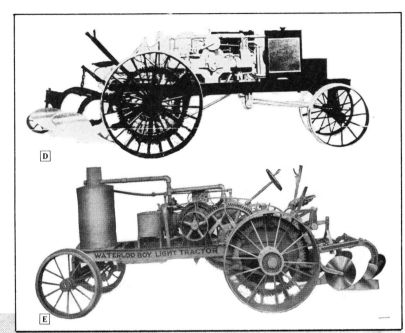

The Waterloo Boy "R"—1914

The first four styles, A to D, of the new Model "R" had 5½″ × 7″ engines, and consisted of 115 tractors: 18 style A, 29 B, only 3 of C, and 65 of D. The cylinder size was then increased to 6″ × 7″ and production began in earnest.

212 style E tractors were built and one or two have survived with collectors. It was with the next style—F—that the company entered the export market. With the world at war, No. 1407 was the first of this model shipped to Britain to help combat the U-boat menace by increasing food production. Subsequently some 4000 Waterloo Boys of this and later models were sent to the British Isles.

These tractors were imported by L.J. Martin of London, who, because of the extra effort required at that time, had called his company the Overtime Tractor Co. As a result the British tractors were renamed Overtime and carried decals showing the rear view of a tractor above the face of a clock—"over time." Special serial number plates were fitted, and an example of these is shown on page 97.

With the large demand from the war zone and increasing requirement at home, the production numbers of Waterloo Boy tractors increased rapidly from 1916 to 1918. 193 style F and 104 style G tractors were built, and all these had the original type of engine. With the introduction of style H in 1916, the engine was altered to include a removable cylinder head in place of the cast-in-one version used until then.

A Waterloo Boy Model "C" All-Wheel-Drive tractor fitted with a 2-cylinder opposed 5½″ × 6″ 15-hp motor similar to that of the model "L"

B Waterloo Boy Model "C"—another view

C

D

E Ken Kass's Model "R" No. 1,643 at the sesquicentennial celebrations organized by the Two-Cylinder Club on the company's demonstration grounds at Waterloo, Iowa, in July 1987

F Waterloo Boy No. 1,720 showing the optional vertical fuel tank which was available for a time. The tractor belongs to Don Dufner of Buxton, North Dakota.

G The oldest known example of an Overtime, the British name for Waterloo Boy No. 1,747 at the Wiltshire School of Agriculture's museum at Lacock, near Chippenham, England

F

C Possibly the earliest example in existence of the Waterloo Boy 2-cylinder tractors which laid the foundation of the John Deere line and started the trend to tractor power in World War I is Travis Jorde's Waterloo Boy "R" No. 1,568. With cylinder block and head cast as one, it is fitted with small-diameter fuel tank, round-spoke wheels, and radiator on the right side, and is shown "at home."

D View of No. 1,568's engine

E

G

95

A Model "R-H" No. 2,512 at the Waterloo, Iowa, celebrations in 1987

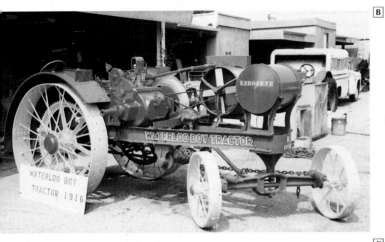

B Model "R-I" No. 3,864 owned by L.W. Grady of Claremont, California

C Steve Just's nicely restored Overtime No. 4,644, brought back by him from England

D A Northern Ireland Overtime, Robert Patterson's No. 4,966. This style L Model "R" is interesting because it was originally sold by Harry Ferguson when he was Overtime agent for L.J. Martin of London in Ireland during the First World War.

F Waterloo Boy decal on "R" No. 4702

G The decal on Steve Just's Overtime

E A good example of the serial number plate fitted to an Overtime tractor

OVERTIME FARM TRACTOR
COMPANY,

124-127 *Minories,*
London, E. 1.

Telephone—AVENUE 5137.
Telegrams—"SUBSOIL, ALD., LONDON"

L. J. Martin, Proprietor.
Address all Communications to the Company.

WE BELIEVE a careful comparison of the detailed design of the "Overtime" Tractor with others of the same class will do more than anything we can say to convince you of the superiority of our machines. This is the most effective and satisfactory way to sell our machines, for it eliminates that feeling of doubt which every one seems to have of statements made by parties concerning their own goods.

HDP/1/LGK

JUNE
27th
1917

Mr. W. Reynolds
Fordfield House
AMPTHILL. Beds.

Dear Sir,

We are obliged by your favour and will again take the matter up with Messrs The Pytchley Autocar Co. with regard to Mr. O.J. Day's Tractor.

We note that the position is now cleared up with regard to Messrs Brown and Co. of Leighton Buzzard. With regard to the Tractor which you bought on conditions that it was delivered from stock, this is quite correct, the machine has been despatched to you from Liverpool and should have reached you ere now. We have made repeated enquiries from our Shippers, and from the information just received, we think that something must have happened to same in transit. At any rate, we have made a claim on the Railway Company for the machine. To save further loss of time, we have today despatched from London, the only machine that we have in stock and trust it will reach you without any great delay.

Assuring you of our best services at all times.

Yours truly,
FOR THE OVERTIME FARM TRACTOR CO.

H. Stoner.

A An original letter
from the Overtime Farm Tractor
Company to their Bedford dealer, W. Reynolds

The
Overtime
FARM TRACTOR

PRICE
£325

B The front cover of an original 1917
Overtime catalog

C An Overtime tractor during World War I
with winner's shield. The writing on the
shield reads, "National Food Production
Campaign—Champion Tractor of England
and Wales." It won by plowing 154 acres in
the four weeks ending March 8, 1917.

D 1915 advertising poster for the Waterloo Gasoline Engine Co.

The 2-Speed "N" Joins the "R"—1917

Some 1,867 style H tractors were followed by 689 style I, 54 K, and 2,501 L. The single-speed "R"'s final style M, introduced in 1917, had a larger 6½″ × 7″ motor, and this same unit was used in the 2-speed "N," introduced shortly afterward. Both models were in production when Deere acquired the company in March 1918.

To complete production details, there were 2,399 style M tractors built, a total of 8,081 Model "R"s in 12 styles over some four years. By comparison some 20,534 units of the 2-speed Model "N" were built during its 7-year life, and a number of these were included in the figure given for shipments to Britain. There are currently over 40 "Overtimes" with collectors in the US and UK.

[B] A picture of a Waterloo Boy "N" tractor in 1917, before Deere's purchase of the company

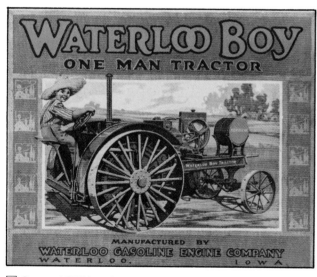

[A] Cover of 1916 brochure describing Waterloo Boy "R" tractors

[C] The only Waterloo Boy "N" tractor recorded so far with a serial number among the "D"s, No. 31,353, belongs to the Sottongs of Tipton, Indiana, and was shown at the Waterloo, Iowa, celebrations in July 1987

D Model "N" with corn binder in the fall of 1918

E Model "N" plowing in June 1920

F Waterloo Boy "N" with road grader

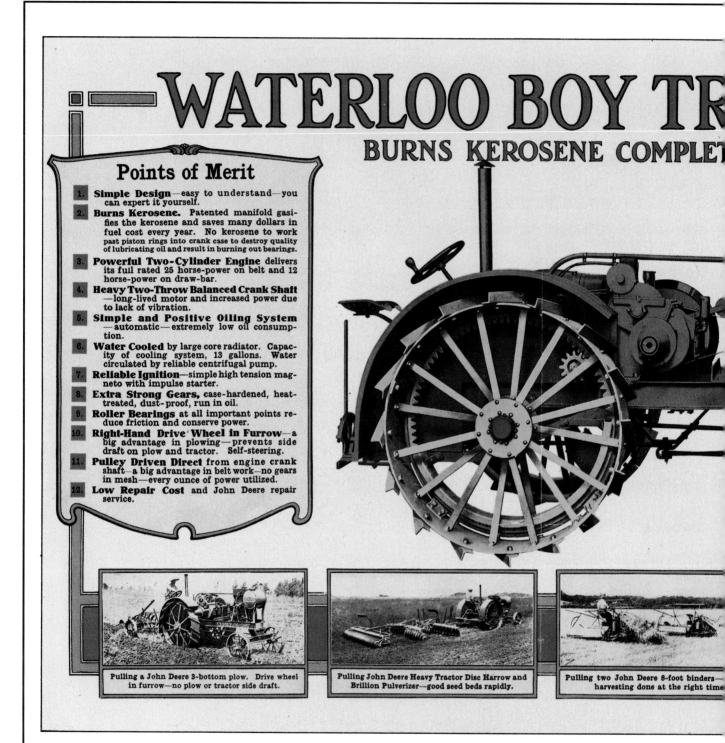

A An advertisement for a later model Waterloo Boy "N" tractor

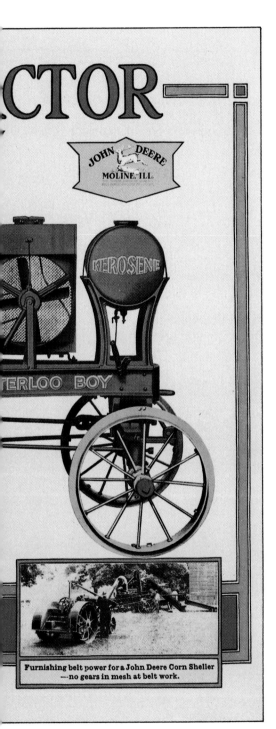

JOHN DEERE
MOLINE, ILL.

KEROSENE

WATERLOO BOY

Furnishing belt power for a John Deere Corn Sheller
—no gears in mesh at belt work.

B An unusual overhead view of a Waterloo Boy "N" before they were modified with automobile-type steering

Deere Commences Tractor Experiments—1912

While the Waterloo factory had grown to achieve this sort of output, Deere engineers were experimenting with tractors themselves. The year 1912 saw the first of these units designed by C.H. Melvin, and then in 1915 Joseph Dain produced his first model, followed by three further improved types. These were 25-hp all-wheel-drive tractors. When units of the final type were fully tested, they acquitted themselves very well, so a production run of 100 was built in 1918-19.

Soon after the purchase of the Waterloo company, Deere continued experiments to build a tractor with totally enclosed final drives. Several interesting pictures of some of these tractors are shown on page 107, the earlier models still called Waterloo Boy.

B

C

A Deere's first experimental tractor—the Melvin of 1913-14

D

B The Melvin tractor and its plow

C The production All-Wheel-Drive tractor with spring-tine harrow

D Detail of production All-Wheel-Drive tractor's engine

E Four Dain tractors with four implements made in the Marseilles (Spreader) Works

F Possibly the last example of the All-Wheel-Drive tractor, with two rows of hood vents and low-down shielding between the engine compartment and the fuel tank

E

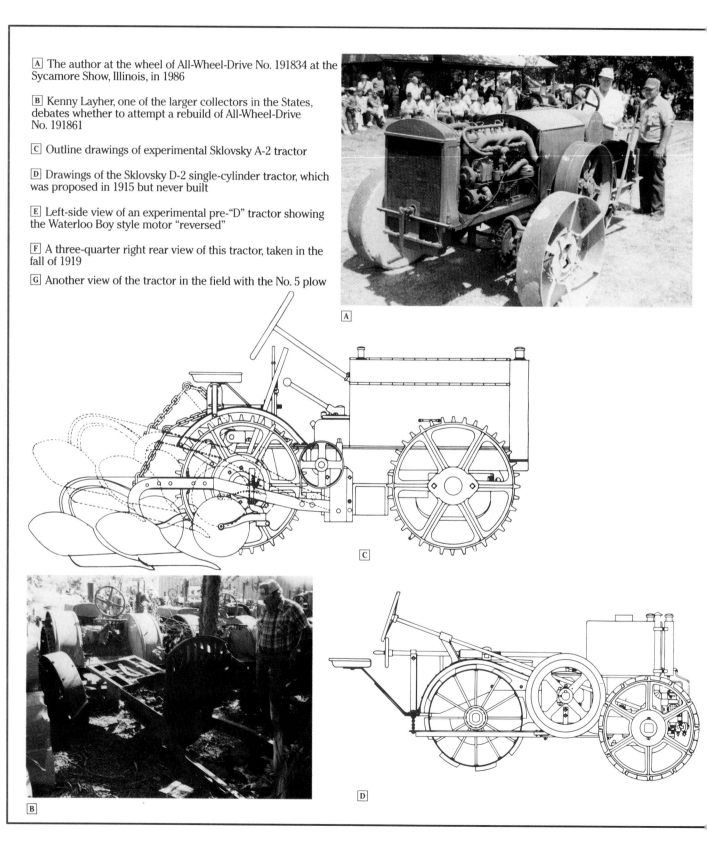

A The author at the wheel of All-Wheel-Drive No. 191834 at the Sycamore Show, Illinois, in 1986

B Kenny Layher, one of the larger collectors in the States, debates whether to attempt a rebuild of All-Wheel-Drive No. 191861

C Outline drawings of experimental Sklovsky A-2 tractor

D Drawings of the Sklovsky D-2 single-cylinder tractor, which was proposed in 1915 but never built

E Left-side view of an experimental pre-"D" tractor showing the Waterloo Boy style motor "reversed"

F A three-quarter right rear view of this tractor, taken in the fall of 1919

G Another view of the tractor in the field with the No. 5 plow

A

C

B

D

E

F

G

107

Arrival of the Model "D"—1923

The year 1923 signalled the arrival of the famous Model "D" on the market—the shape of things to come for Deere, and the model which achieved widespread farmer acceptance for the next 30 years. Used all over the world, the "D" started life with a 26″ spoke flywheel, direct steering from wheel to front axle, and left-side steering—unlike its predecessor, which had always had a right-side driver's position.

It soon became obvious that the steering rod was too close to the flywheel for comfort—and safety—so a jointed steering rod which had been tried on some of the experimental tractors was substituted. At the same time the flywheel was reduced in diameter to 24″, but increased in width to give the same weight. Some 879 of the 26″ type were built in 1923-24 before the change, and 4,876 with the 24″ in 1924-25.

For the 1926 season the flywheel fitted was solid cast, but still with a key to the crankshaft as with the spoke type. The following year saw the introduction of the splined crankshaft and flywheel, and at the same time the engine bore was increased from 6.5″ to 6.75″, although the 2-speed transmission was retained. The biggest change to the "D" came in 1931 with the alteration of right-side worm-and-gear steering, and an increase in the governed engine speed from 800 to 900 rpm.

It was not until 1935 that a 3-speed transmission was added to give the "D" its final mechanical form. Styling was added in 1939 to bring the model into line with its row-crop counterparts.

A A rear view of a 1922 pre-"D" tractor

B One of the first "D"s with a binder in oats

C The Model "D" started a new era with a line of simple, sturdy tractors the ordinary farmer could use and maintain. Its horizontal 2-cylinder concept gave the company its unique identity for many years to come. The photograph shows one of the first 50 "D"s plowing with a No. 5 plow.

D A 1924 "D" in California

F

G

E One of the first 50 Model "D" tractors at work with a Sheldon Triplex road scraper

F The tenth "D" made, now owned by the Layhers of Wood River, Nebraska, in the process of restoration on Sept. 9, 1986

G A front view on the same day showing the fabricated front axle of the first 50 "D"s

H An early attempt at an industrial version of the "D"

E

H

A

B

C

JOHN DEERE MODEL D TRACTOR WITH SPEED REDUCER

D

A The only 26″-spoke "D" in England, sold on May 2, 1987, due to the death of Eric Armistead, its original importer and a well-respected collector

B A beautifully restored "D" with 24″ spoke flywheel, owned by the Bellins of Isanti, Minnesota. Son Mike does the restoring.

C An option available in 1930 was a speed reducer for the "D"

D The controls necessary to equip a "D" and binder for one-man operation

E "D" No. 43,479 in a sale at Sully, Iowa, on July 31, 1987

F A statement of company policy in 1937, after many years' experience

E

A FOUR-CYLINDER ENGINE—NOT ON YOUR LIFE

Some of our dealer and tractor owner friends occasionally tell us they have heard that John Deere is coming out with a four-cylinder engine in all John Deere Tractors. Dealers and others who are well informed know that this rumor is false.

We want to assure those who are not so well informed that we are *NOT* coming out with a four-cylinder engine. *THE JOHN DEERE TWO-CYLINDER ENGINE HAS BEEN SO OUTSTANDINGLY SUCCESSFUL THAT THERE IS NO THOUGHT OF A CHANGE.*

JOHN DEERE TRACTOR COMPANY
L. A. Rowland
Vice Pres. and Gen'l Manager

F

111

Different "D" Applications

Variations on the standard tractor occurred over the years. The first was due to the recognition that it had an industrial application in addition to its traditional agricultural role. Over the years this version, the "DI," was developed by providing industrial-style wheel equipment, braking, and specialized seating arrangements, and the ability to fit industrial machines.

Other variants were a low-built version for orchard work, the "DO," and in the 1930s experiments with crawler half-tracks. In the West, Lindeman fitted three "D"s with full tracks, but these were not satisfactory and all three were scrapped.

Finally the luxury of electric starting and lighting became more common as an optional extra on the "D" in the late forties, when all the other models in the line were so fitted as standard.

B Orchard version of the "D," No. 114,552, shown by the Layhers at the Waterloo function. Note the special air intake.

A Deere experimented with crawler adaptation of the Model "D." It is thought that this is one of those tractors.

E Model "D" No. 104,411 at a sale in Sully, Iowa, on the last day of July 1987

C View of the same tractor from the flywheel side

D Rear view showing the low seating position

F Model "D" fitted with Lindeman crawler tracks, plowing on October 20, 1932

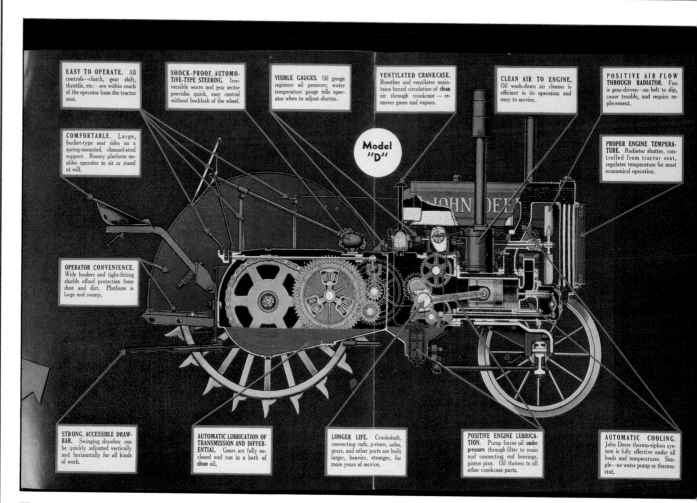

EASY TO OPERATE. All controls—clutch, gear shift, throttle, etc.—are within reach of the operator from the tractor seat.

SHOCK-PROOF, AUTOMOTIVE-TYPE STEERING. Irreversible worm and gear sector provides quick, easy control without backlash of the wheel.

VISIBLE GAUGES. Oil gauge registers oil pressure; water temperature gauge tells operator when to adjust shutter.

VENTILATED CRANKCASE. Breather and ventilator maintains forced circulation of **clean** air through crankcase — removes gases and vapors.

CLEAN AIR TO ENGINE. Oil wash-down air cleaner is efficient in its operation and easy to service.

POSITIVE AIR FLOW THROUGH RADIATOR. Fan is gear-driven—no belt to slip, cause trouble, and require replacement.

COMFORTABLE. Large, bucket-type seat rides on a spring-mounted, channel-steel support. Roomy platform enables operator to sit or stand at will.

Model "D"

PROPER ENGINE TEMPERATURE. Radiator shutter, controlled from tractor seat, regulates temperature for most economical operation.

OPERATOR CONVENIENCE. Wide fenders and tight-fitting shields afford protection from dust and dirt. Platform is large and roomy.

STRONG, ACCESSIBLE DRAWBAR. Swinging drawbar can be quickly adjusted vertically and horizontally for all kinds of work.

AUTOMATIC LUBRICATION OF TRANSMISSION AND DIFFERENTIAL. Gears are fully enclosed and run in a bath of clean oil.

LONGER LIFE. Crankshaft, connecting rods, pistons, axles, gears, and other parts are built larger, heavier, stronger, for more years of service.

POSITIVE ENGINE LUBRICATION. Pump forces oil **under pressure** through filter to main and connecting rod bearings, piston pins. Oil thrown to all other crankcase parts.

AUTOMATIC COOLING. John Deere thermo-siphon system is fully effective under all loads and temperatures. Simple—no water pump or thermostat.

A Cutaway of Model "D" (side view)

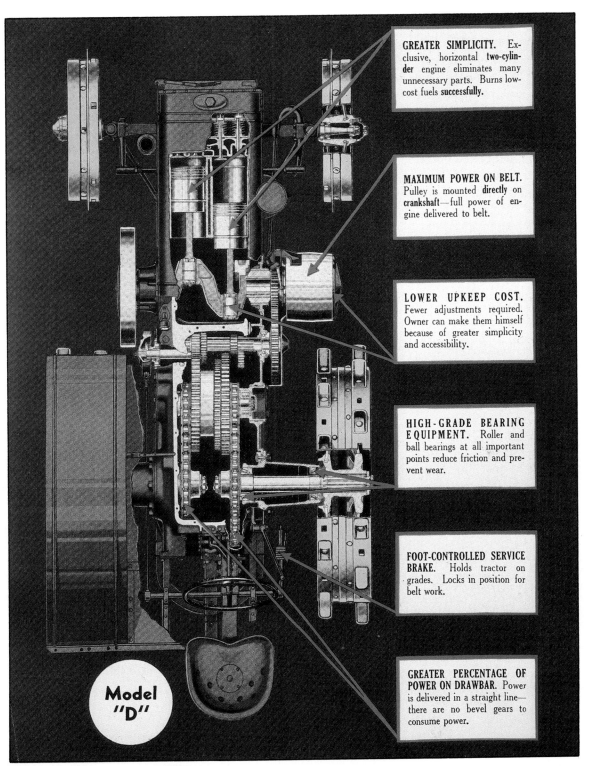

GREATER SIMPLICITY. Exclusive, horizontal **two-cylinder** engine eliminates many unnecessary parts. Burns low-cost fuels **successfully.**

MAXIMUM POWER ON BELT. Pulley is mounted **directly** on **crankshaft**—full power of engine delivered to belt.

LOWER UPKEEP COST. Fewer adjustments required. Owner can make them himself because of greater simplicity and accessibility.

HIGH-GRADE BEARING EQUIPMENT. Roller and ball bearings at all important points reduce friction and prevent wear.

FOOT-CONTROLLED SERVICE BRAKE. Holds tractor on grades. Locks in position for belt work.

GREATER PERCENTAGE OF POWER ON DRAWBAR. Power is delivered in a straight line—there are no bevel gears to consume power.

Model "D"

B Cutaway of Model "D" (overhead)

For Row-Crop Farmers, the "C" Arrives—1927

Although experiments on motor cultivators had been abandoned in 1921, the row-crop application of tractors was still very much on the agenda. In 1926 a design was developed for a smaller tractor with row crops in mind, and five tractors were built. The next year production started on a further 24 machines, followed by 75 more designated the Model "C."

Various teething troubles caused the withdrawal of all of them for modification, and 52 were rebuilt, renumbered, and reissued, with 23 further new machines added, all numbered between 200,111 and 200,186.

B Closeup of a Model "C" fitted with cultivator in April 1927

A The first tractor to provide power from four outlets—drawbar, belt pulley, power takeoff, and the power-driven mechanical lift for integral equipment. The picture shows a well-restored Model "C," No. 200,109, belonging to the Kellers of Forest Junction, Wisconsin.

C Studio photograph of Model "C"

D Model "C" and 3-row corn planter at Marsh Farm, Waterloo, Iowa, in May 1927. Note the unusual front axle, left-side steering rod, and no name on radiator top tank.

E Rear view of Model "C" with 3-row planter in May 1927. Note wider front wheels but intermediate-size rear wheels.

F Tricycle "C" and cultivator in June 1928

"C" Renamed "GP" and Production Begins—1928

A new series of "GP" or General Purpose tractor, starting with No. 200,211, was announced in 1928. The change in model letter from "C" to "GP" was variously ascribed to the need to compete with International Harvester Company's Farmall or F series, or to avoid the similarity in sound of "C" and "D." Whichever was the case, the Deere tractors retained the prefix "C" for all their part numbers, and the original batch kept the "C" on their serial number plates.

[B] An ingenious representation of a "GP" Tricycle tractor, assembled and owned by Travis Jorde of Rochester, Minnesota

[A] The Bellins' Model "C" No. 200,167 at Isanti, Minnesota

[C] The company's restored "GP" No. 205,205 on the Moline display floor on October 6, 1986

D "GP" No. 210,113 was one of four tractors raffled at the 150th year celebrations in Waterloo, Iowa, July 1987

E Scott Zoborosky's 1931 "GP" No. 228,505 shows this model in its latest form apart from minor detail changes

A Tricycle Tractor Built for 2- and 4-Row Work—1929

Of the original five pre-production machines, the first and last had been tricycle types, the second with wide tread. It is not surprising, therefore, to find that some 14 tricycle tractors built in 1928-29 were numbered among the standard "GP"s. A list of these appears in the appendices, and one known to have survived with a collector is No. 204,213, the property of Walter and Bruce Keller of Forest Junction, Wisconsin. This tractor is pictured as restored by them and shown at the Waterloo show. One other tractor of similar design, but made up from the front of a wide-tread and the rear of a standard tractor, was also exhibited.

The natural development from this exercise was the introduction of the wide-tread model, known today as the "GPWT." Originally it had the air cleaner under the hood, and had the stub exhaust of the standard model. Subsequent series had the air-cleaner intake through the hood, and after the introduction of the larger 6″ × 6″ engine the exhaust and air intake changed sides and both were external to the hood.

B Front view of the only known "GP" Tricycle to have survived—No. 204,213, the property of the Kellers of Forest Junction, Wisconsin

C Another view of this tractor, with Bruce Keller at the wheel

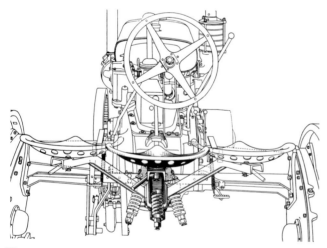

A An optional extra for the "GPWT" tractors from 1931 on was this swinging seat for use in row-crop work

D "GPWT" No. 400,047, the property of the Layhers of Wood River, Nebraska

E An early model "GPWT," No. 402,204

F Another view of No. 402,204

The Shape of Row-Crop Tractors to Come—1932

A final further series appeared in 1932 with steering over the top of the much slimmer tapered hood, which in turn gave greatly improved visibility for the operator. This represented a preview of the shape of the next row-crop design to be announced, of which more shortly. These overhead-steer "GPWT" tractors are in great demand with collectors, and command high prices as a result.

B "GPWT," No. 400,047 owned by the Layhers of Wood River, Nebraska. Originally numbered as a standard, this is an example of one of the first "GPWT"s built after the Tricycle tractors.

A A rear view of a late wide tread showing the high clearance available with these models. The tractor was at the Farmfest Show at Lake Crystal, Minnesota, held to celebrate the Bicentennial of the USA.

C "GPWT" No. 402,204

D A line-up of nicely restored tractors at the Bellins' in Isanti, Minnesota. The overhead steer "GPWT" is No. 404,812 with a GPWT side-steer, "GPO," and "AI" in the background.

E GPWT No. 405,178 at a vintage show at Marshalltown, Iowa on August 1st, 1987.

F "GPWT" No. 405,190 fitted with inset potato rear wheels giving a 68″ tread. These wheels were introduced to fit the standard wide-tread models in place of the special quills fitted to the earlier Series "P" tractors.

A Model for the Potato Grower

Prior to these developments, in 1930 a wide-tread of 68″ width to fit the potato rows in Maine had been offered and was called the Series "P." The first 150 tractors were built new, but a further 53 were rebuilt from standard machines. At least 36 of these are known to have survived, at the time of writing, and others will doubtless be found deep in Maine, where they were all shipped.

A From Isanti, Minnesota, the Bellins' Series "P" No. 5,147 with their "C" No. 200,167 in the background

B Illustration of a Series "P" from advertising material

C Series "P" No. 5,165 on rubber tires and
Series "P" No. 5,164 on steel with mid-
mounted cultivators, owned by the Coblers
of Ottumwa, Iowa

An Crawler and Orchard Version

Other variations on the "GP" theme included the introduction in 1930 of an orchard model, the "GPO." Some of these were in turn purchased by Lindeman of Yakima, Washington, and fitted with full crawler tracks. Eleven of some 25 tractors so adapted have survived to the present day. Although not appreciated at the time, this could be seen as the starting point of the company's crawler tractor interest, which led eventually to the worldwide industrial line available today.

B Rear view of No. 15,703 opposite

A The orchard version of the "GP," No. 15,149, at a sale near Sully, Iowa, on July 31, 1987

C Rear view of No. 15,323 shown opposite

D A rare "GPO"-Lindeman, No. 15,703 owned by John Nikodym of Red Cloud, Nebraska.

E A "GPO" No. 15,323 fitted with full citrus fenders. Property of Lloyd Scheffler of Mount Auburn, Iowa.

The Model "A" Row-Crop— First with Hydraulics

During 1932, after the introduction of the last style "GPWT," experiments were proceeding on a more advanced row-crop model to take its place. The FX and GX experimental models were the links to this new model, originally designated "AA." Initially six experimental tractors were built in April 1933, four with 4-speed transmissions, the AA-1, and two with a 3-speed verison, the AA-3. Two more 4-speed were built in June, at which time the 3-speed option was cancelled. Four further tractors were built that year, but all 12 were returned to the Works, 10 to be rebuilt to conform to the production model. One of the 3-speed was scrapped, and the fate of the other unit is not known.

An appendix lists these tractors and their subsequent serial numbers when rebuilt. Production of the new "A" row-crop tractor, with splined rear axle for ease of wheel adjustment, hydraulic lift (a first on any tractor), excellent driver visibility, and 4-speed transmission, commenced in April 1934. The first 4800 built had an open driveshaft to the fan, and all these were twin-front-wheel models. In 1935, starting with No. 414,809, this shaft was enclosed, as it always was on the smaller Model "B" announced late in 1934.

The "A"s will go down in farm tractor history as the first to apply hydraulics in place of the earlier hand lifts and more recent mechanical lift of the "GP"s.

Ⓐ Model "AA" tractor with 4-row cultivator shown working in a field in Arlington, Texas, June 2, 1933

Ⓑ Photo of experimental tractor leading to the development of the Model "A"

C Model "AA-1" experimental tractor with GPA 472 cultivator

D Model "AA-1" with GPA 102 bedder, January 1934

E Open-fanshaft "A" No. 411,837 with John Deere thresher at the Waterloo show

F An example of the "A" with open fanshaft, No. 411,880, at the Sully, Iowa, sale in July 1987

Front and Rear Axle Options

The options of a single-front-wheel "AN" and wide-adjustable-front-axle "AW" were also introduced for the 1935 season. It was two years later that models with greater clearance, the "ANH" and "AWH," were added to the line. To obtain this requirement, the tractors were fitted with 40″ rear wheel equipment in place of the standard 36″, and the single-front-wheel tractors had 16″ instead of 10″ fronts.

B Consecutive serial numbers on an "AN" and "AW" at the Waterloo show—Nos. 429,727 and 429,728

A "AW" No. 422,548, for sale at Sigourney, Iowa, on August 1, 1987

C "AWH" No. 472,018, belonging to Bruce Aldo of Westfield, Massachusetts

D Walter Keller's unstyled "ANH" No. 475,074 at Waterloo

E Group of Model "A" tractors:
Front—"AOS" No. 1,457, "AW" No. 429,728, "AN" No. 429,727.
Rear—"AW" 4-speed, "AN" late, "A" late, "AI," "AWH" No. 472,018, unstyled "A"s on rubber and steel

Replacing Two Horses, Enter the "B"— 1934

As in the Waterloo Boy days, and again in the twenties, when the "D" was the only model in the line, the demand again arose for a tractor smaller than the "A." The result for the 1935 season was the introduction of the Model "B," designed to replace two horses on the farm.

Similar in most respects to its bigger brother, it was initially introduced with a shorter main frame, but from No. 42,200 on this was lengthened so that integral equipment could be interchanged between the two models. Again the single-front-wheel "BN" and wide-axle "BW" options were available, and the high-crop versions, the "BNH" and "BWH," were announced in 1937 with their "A" counterparts.

B Another of the first style of the Model "B"

A The first 42,200 production "B" tractors had a short frame, as in the one illustrated here

C A very original short-frame "BW," No. 37,434, at Lloyd Bellin's at Isanti, Minnesota, in 1986

D Studio picture of unstyled "BN" on rubber

E "BW" No. 57,351, brought to Waterloo by Bruce Johnson of South Elgin, Illinois

A

B

C

A Unstyled "BN" of the later type, on steel wheels

B Lloyd Sheffler's short-frame "BN" No. 17,149 at Waterloo

C BN-Garden tractor photographed with Wendell Helphrey's 1935 "B" No. 7,593 from Winfield, Iowa

D "BWH-40," owned by Don Dufner of Buxton, North Dakota

E The same tractor at the farm of the previous owner, Ken Berns of Blue Hill, Nebraska

F Front view of this unique tractor shows its very narrow lines.

D

E

F

Smaller Standard 4-Wheel Models Introduced—1935

To satisfy the demand of those who wanted standard tread tractors smaller than the "D," the "AR" and "BR" (regular) models made their debut in 1935. Also announced at the same time were orchard versions with lower profile, independent rear wheel brakes for shorter turning among the trees, and the option of enclosed rear wheel fenders.

The following year and until 1941, adaptations of the "AR" and "BR" were produced for industrial use. The "AI" and "BI" were similar in appearance to the regular models, but had drilled plates added to the main frames for the easy attaching of blades, sweeps, and other industrial units. In addition they were usually fitted with downswept exhaust and low air intake similar to the orchard models, industrial-type heavy-duty hand-operated rear wheel brakes, and upholstered seat with backrest.

B An early Model "AO" with cast offset radiator cap, as supplied prior to their streamlining. Serial numbers of this style were between 250,075 and 253,482.

A The standard-tread version of the "A" was well represented at the Waterloo show. This one, No. 257,918, is owned by Emi Kuntz of Grafton, Iowa.

C Standard-tread "BR" on French & Hecht wheels, owned by Robert Dufel, seen here with his father and his wife, Shirley

D "BR" No. 331,998 on steel sold for £3,750 ($6,500) in a sale in England in May 1987

E Another "BR," No. 332,004, owned by John Deere dealer Jim Dance of Princes Risborough, England

F Very popular in England are the Model "BR"s. This one, No. 330,487, is at a Langport, Somerset, show.

A A "BI" stands in front of the Layhers' farm house at Wood River, Nebraska

B The "AI" in industrial yellow, shown at Waterloo, belonged to Ed Brenner of Kensington, Ohio

C Loading up the tractors owned by Rodney and Emi Kuntz after the Waterloo show. "AR" No. 257,918 and "B" No. 1,007 safely aboard; "BR" No. 325,516 on steel wheels mounts the ramps.

D A Model "AI" used for many years in the Waterloo Works

E "BI" No. 327,568, good looking in its industrial yellow color scheme. The property of Terry Ploughe, Tipton, Indiana

F Models "BI," "AI," and "DI" at the Layhers', Wood River, Nebraska

G A rear view of DI No. 127,364 owned by the Bellins of Isanti, Minnesota, showing the offset driver's position

A Streamlined Orchard Model Announced—1936

From 1936 to 1940 the "AO" was streamlined to become the "AOS," in today's collector parlance. At the same time a separate batch of serial numbers between 1000 and 1891 was allocated to this model.

When the "AR" and "AO" were given larger engines with an extra ¼" stroke (5½" × 6¾" instead of 5½" × 6½"), the streamline version of the AO was dropped, and both models acquired a centered pressed-steel radiator cap in place of the offset oval cast type used until then. It was a brand new "AR" of the earlier type, No. 259,257, delivered to a Gloucestershire farm where the author worked, that introduced him to "John Deere, Moline, Ill." and resulted in the green and yellow blood now flowing in his veins! The new models had serial numbers from 260,000 up.

Some 2000 of the "BO" tractors were shipped to Lindeman at Yakima, Washington, to have their crawler track conversion fitted to meet a popular West Coast requirement. These tractors are shown on page 142.

B Another "AOS," No. 1,239, at a sale at Sigourney, Iowa, in 1987

A Robert Pollock of Denison, Iowa, brought this beautifully restored "AO" Streamline tractor, No. 1,457, to the Waterloo show

C "DI" No. 127,364 from the Bellins of Isanti, Minnesota. Fewer than 100 of this model were produced.

D Provided by James Stewart of Rushville, Indiana, the standard "D" No. 128,430 makes an interesting comparison with the DI opposite.

E "AR" No. 261,720 with center pressed-steel radiator cap and the larger 5½″ × 6¾″ engine introduced in 1941, seen at Devizes, England, in 1981

A "BO"-Lindeman No. 326,225 at Waterloo; owner, Steve Stevenson of Bartlett, Illinois

B A very late example of the "BO"-Lindeman crawlers. Note the hydraulic controls, which were an optional extra. No. 337,336 was in the Sigourney, Iowa, sale on August 1, 1987.

C A "BO"-Lindeman, No. 334,592, at the Sully sale on July 31, 1987

D A "BO"-Lindeman, No. 334,558, with its owner at the controls—Mr. Van der Hart of Pella, Iowa

E Lee and Betty Norton's "AO" arrived at the Waterloo, Iowa, show from Alto, Michigan.

F Pulley-side view of the same tractor

The Line Styled for Tomorrow—1938

The year 1938 saw the introduction of styling to the "A" and "B" series row-crop tractors. Initially and for nearly three years, these new models retained the 4-speed gearbox of their predecessors, but in 1941 a 6-speed transmission became standard. This was accomplished with a 2-lever shift; one lever gave three forward speeds and reverse while the second gave Hi-Lo options in all gears.

In the case of the "A," the larger engine was fitted at this time, as referred to above. The "B" series had received their increase when the row-crop models were styled—from $4\frac{1}{4}'' \times 5\frac{1}{4}''$ to $4\frac{1}{2}'' \times 5\frac{1}{2}''$. Governed engine speed remained the same for the new engine sizes on both models.

Despite the styling of the row-crop tractors, the standard and orchard models had to wait until 1949 for their new look—a delay possibly due to the intervention of the war.

Ⓐ A 1940 4-speed "AW" owned by Robert Tucker of Rushville, Indiana

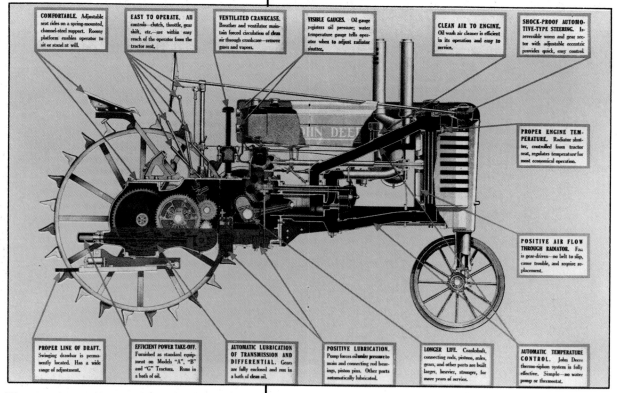

COMFORTABLE. Adjustable seat rides on a spring-mounted, channel-steel support. Roomy platform enables operator to sit or stand at will.

EASY TO OPERATE. All controls—clutch, throttle, gear shift, etc.—are within easy reach of the operator from the tractor seat.

VENTILATED CRANKCASE. Breather and ventilator maintain forced circulation of clean air through crankcase—remove gases and vapors.

VISIBLE GAUGES. Oil gauge registers oil pressure; water temperature gauge tells operator when to adjust radiator shutter.

CLEAN AIR TO ENGINE. Oil wash air cleaner is efficient in its operation and easy to service.

SHOCK-PROOF AUTOMO-TIVE-TYPE STEERING. Irreversible worm and gear sector with adjustable eccentric provides quick, easy control.

PROPER ENGINE TEMPERATURE. Radiator shutter, controlled from tractor seat, regulates temperature for most economical operation.

POSITIVE AIR FLOW THROUGH RADIATOR. Fan is gear-driven—no belt to slip, cause trouble, and require replacement.

PROPER LINE OF DRAFT. Swinging drawbar is permanently located. Has a wide range of adjustment.

EFFICIENT POWER TAKE-OFF. Furnished as standard equipment on Models "A," "B" and "G" Tractors. Runs in a bath of oil.

AUTOMATIC LUBRICATION OF TRANSMISSION AND DIFFERENTIAL. Gears are fully enclosed and run in a bath of clean oil.

POSITIVE LUBRICATION. Pump forces oil under pressure to main and connecting rod bearings, piston pins. Other parts automatically lubricated.

LONGER LIFE. Crankshaft, connecting rods, pistons, axles, gears, and other parts are built larger, heavier, stronger, for more years of service.

AUTOMATIC TEMPERATURE CONTROL. John Deere thermo-siphon system is fully effective. Simple—no water pump or thermostat.

Ⓑ Cutaway of 4-speed "A"

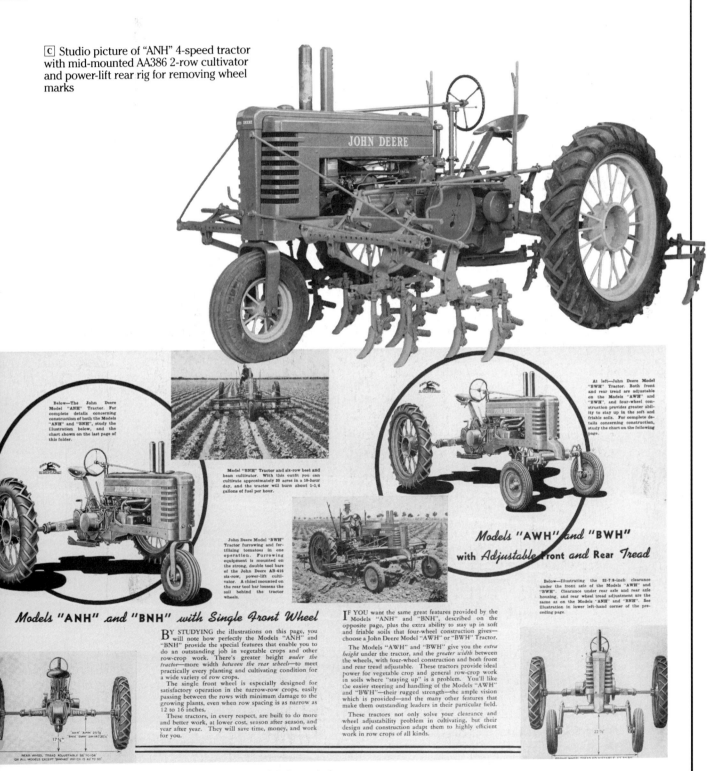

C Studio picture of "ANH" 4-speed tractor with mid-mounted AA386 2-row cultivator and power-lift rear rig for removing wheel marks

D Original sales literature on high-crop models for styled era

A Styled 4-speed model "BWH"—studio picture

B Single-front-wheel version of the "A"—the "AN"—from Northwood, Iowa, owned by Mike Stevens

C Photograph of 6-speed "A" owned by Alvie Grossman Jr., Pryor, Oklahoma, and a similar model on rubber belonging to Dennis Bensheimer

D "BW" No. 60,617 provides transport from church to reception for the bride and groom, Dr. Geoff and Veronica Mathews of New House Farm, Llangwm, Usk, Wales

E Model "A" 6-speed with optional electric starting and lighting at a vintage show organized by the Rotary Club of Boswell, Indiana, in 1984

F A nicely restored "BN," No. 122,681, in the workshop of Mike Towler at St. Germans, Norfolk, England, in 1987

G "BW" No. 155,639 at a sale in Hampshire, England, in 1987. The numbers painted on its front pedestal are the U.K. road registration number required for all self-propelled vehicles.

H "AW" No. 531,093, shipped from the factory on December 17, 1943, to England. Having survived the Atlantic crossing it is seen, duly restored, on a farm in Oxfordshire in 1985.

Farmers Demand a Larger Row-Crop Tractor—1938

With the ever-increasing power requirements of the row-crop farmer, a tractor nearer to the "D" in power was needed, and in 1938 the "G" was introduced in unstyled form. It was originally intended to call this model the "F," but confusion with International Harvester's F series caused a change of mind. Nevertheless, all the parts for the "G"s retained their original "F" prefix.

Initially fitted with a radiator which was rather too small in some conditions—4500 tractors were produced before a decision to increase the size of this item was made—the "G" had a 4-speed gearbox and was the largest row-crop tractor on the market. Its 6⅛" × 7" engine operated at the same governed speed as the "A"—975 rpm.

Over 11,000 units were produced from 1938 to 1941, when the company decided to style the model and give it a 6-speed transmission like the new "A" and "B" models. The reason for the change in its model letter has already been explained. The additional M in "GM" may have been for "modernized" or "modified," but only lasted as long as restrictions so demanded. With this update a single front wheel and a wide front axle were offered as options. These had not been available on the earlier model.

B A group picture of "G"s at the Waterloo show covering this model's history. From lower right counterclockwise, the low-radiator "G" No. 1,006, large-radiator "G" No. 8,373, "GM" No. 14,074, "GH" No. 47,680, and the last "G" built, No. 64,530

A A restored "G" on skeleton rear wheels and cast disk fronts. No. 11,673, at the July 31, 1987, sale at Sully, Iowa

C Low-radiator type "G" No. 2,810, originally No. 1,000, the first built. Owned by George Morse of Austin, Minnesota

D   E

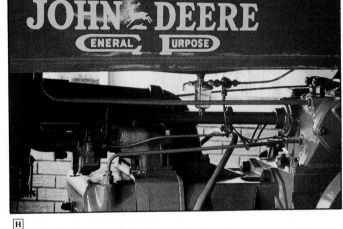

F   G   H

D Murray Simpson of Cupar, Fife, Scotland, owns this beautifully restored Model "G," No. 5,822, seen here in his Ceres dealership showroom.

E Flywheel side of the same tractor

F Close-up of the single-lever gearshift adopted first for the "G" tractors

G The French & Hecht rear wheel of the same tractor

H Decal and engine detail in close-up of "G" No. 5,822

I Illustration from sales literature for Model "GM"

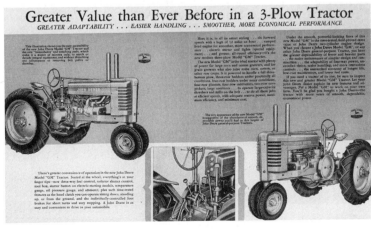

Greater Value than Ever Before in a 3-Plow Tractor
GREATER ADAPTABILITY . . . EASIER HANDLING . . . SMOOTHER, MORE ECONOMICAL PERFORMANCE

I

An 8-hp Baby for the Line

Back in 1936 the Wagon Works had experimented with an 8-hp baby tractor, designated the Model "Y," and 24 of these units had been built, fitted with a Novo 2-cylinder vertical gasoline engine. Resulting from the success of these small tractors, a production run of an updated version, the Model "62," was built in the summer of 1937, and a further series of nearly 4000 tractors, again slightly modified and renamed the "L," was built in 1937 and 1938.

These tractors introduced the idea, much copied later, of an offset (left side) engine and transmission, with the operator offset slightly to the left to allow better vision for cultivating and other operations.

Following the enthusiastic customer acceptance of the larger styled tractors, the "L" was similarly treated in 1939, and an industrial version, the "LI," was added to the line. In 1940 a slightly larger version, the "LA," with another 4 hp—14.34 against 10.32—and 24″ instead of 22″ rear wheels, was introduced. Electric starting and lighting were options for both models from this time.

B The requirement for a utility tractor resulted in the production of the experimental Model "Y," built in the Wagon Works in 1936. Shown is a studio picture of one of these units.

A A reproduction Model "Y" built with an original Novo engine by Jack Kreeger of Omaha, Nebraska

C Studio photograph of the original Model "62" tractor

D   E

D Rear view of Model "62" showing the JD on the rear axle casting

E A well-restored "62," No. 621,048, the property of Ron Jungmeyer, Russellville, Missouri, showing the JD on the lower radiator casting

F A group of tractors on display at the Waterloo, Iowa, show representing the "L" family. Clockwise from lower left, "62" No. 1,048, unstyled "L" No. 622,708, "L" No. 631,720 with disk harrow, "LI" No. 51,253 and mower, and "LA" without serial number plate with single-furrow two-way plow.

F

A

B

C

D

E

F

A The industrial model was based on the motor of the "L" and called the "LI." Edwin Brenner's tractor No. 51,253 is shown with a mid-mounted mower

B This unstyled "L" has extra mud lugs, an unusual option

C An unstyled Model "L," no serial number plate, at the Sully, Iowa, sale in July 1987

D Walter Keller leans on his "62" in the early 1980s at the head of some of his collection

E Studio photograph of late-type "L" (after No. 640,000) fitted with John Deere engine in place of the Hercules-built unit on previous tractors

F Well-restored Model "L" No. 629,882 at the Waterloo, Iowa, Show, brought by Jack Kreeger

G Model "LA" No. 5,115 in a sale at Sigourney, Iowa, in 1984

H "LA" No. 11,690, on the Moline display floor, August 3, 1987. This model was a more powerful version of the "L," and both continued in production until 1946. The "LA" with a gas engine produced the same belt horsepower as the all-fuel "H."

I Details from an "LI" brochure

J Front page of the sales brochure describing the model "LI" and its matched equipment

G H

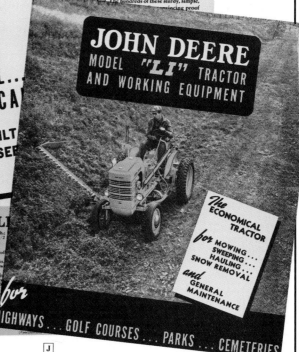

JOHN DEERE Model "LI"

BEFORE BUYING ANY TRACTOR . . .
CONSIDER THIS PRACTICAL COMBINATION OF FEATURES

(1) OUTSTANDING SIMPLICITY . . . only a John Deere gives you the outstanding simplicity of two-cylinder engine design.

(2) OUTSTANDING 3-WAY ECONOMY . . . low purchase price . . . low daily fuel costs (burns only half-gallon of fuel per hour under ordinary conditions) . . . low maintenance costs because of its simplicity and sturdy construction.

(3) EASY TO UNDERSTAND—EASY TO HANDLE. nothing to learn—drives just like an automobile with standard gear shift and foot clutch. Quick-moving. Flexible. Independent foot brakes control rear wheels—the "LI" can be turned in a 7-foot radius. Owners report it to be the handiest, most "sure-footed" power outfit they have used for work in close quarters, sharp turns, and steep ditches.

(4) GREATER DEPENDABILITY . . . fewer, stronger, longer-lived parts made possible by exclusive two-cylinder design.

(5) LONGER LIFE . . . you can expect extra years of service because of high quality of workmanship and use of high-grade materials throughout.

(6) THREE FORWARD SPEEDS . . . the right speed for economical handling of every job.

(7) POWERFUL FOR ITS SIZE—ADAPTABLE TO VARIETY OF WORK . . . weighs about 1,700 pounds. Compactly built. Plenty of power for pulling mowers, trailers, and implements in summer, snow plows in winter. Rubber tires and light weight enable the "LI" to operate on fine lawns without damage to the turf. Rear wheel tread adjustable.

(8) MUFFLER, TOOL BOX, SICKLE CARRIER, OIL GAUGE AND HIGH SPEED ATTACHMENT . . . regular equipment.

(9) ELECTRIC STARTING AND LIGHTING . . . extra equipment.

SPEEDY . . . STRONG . . . POWERFUL . . . ECONOMICA[L]

A TRACTOR BUILT [FOR] ALL-AROUND SE[RVICE]

MODEL "LI[

TRACTOR SPEEDS: First, 3 m.p.h.; [Second,] Third, 8-1/2 m.p.h.; Reverse, 2-1/2 [m.p.h.; high] speeds of 13 to 16 m.p.h.

ENGINE: 2-cylinder, vertical.
Rated speed: 1,550 r.p.m. control[led by] governor.
Bore: 3-1/4 inches. Stroke: 4 inches.
Ignition: High-tension magneto w[ith] impulse starter.
Carburetor: Fixed load jet with adj[ustable]
Air Cleaner: High-stack, oil-wash [type]
Lubrication: Force-feed pressure. O[il]
Cooling: Thermo-siphon. Water C[apacity]
Fuel Tank Capacity: 8 gal. Fuel, [gasoline]

CLUTCH: Single dry plate, automo[tive]

TRANSMISSION: Selective-type [gears,] forged, cut, and heat-treated.

FINAL DRIVE: Enclosed gear-dri[ve]

BRAKES: Individual foot brakes [pro]vides powerful braking action.

JOHN DEERE
MODEL "LI" TRACTOR AND WORKING EQUIPMENT

The ECONOMICAL TRACTOR

for MOWING . . . SWEEPING HAULING SNOW REMOVAL and GENERAL MAINTENANCE

for [H]IGHWAYS . . . GOLF COURSES . . . PARKS . . . CEMETERIES

THE John Deere Model "LI" Tractor has been a success [. . .] The hundreds of these sturdy, simple, [. . .] [convinc]ing proof

I J

153

A Small Row-Crop Tractor

Despite this extensive line of tractors, the need for a row-crop tractor of less horsepower than the "B" was apparent. The success of the Allis-Chalmers Model B emphasized this point, so in 1939 the company introduced its smallest 2-cylinder horizontal-engine model, the "H," developing 14.84 hp at the belt pulley.

Always styled like the larger "A" and "B" but with a 3-speed gearbox, it retained a hand-operated clutch. In this case the pulley housing the clutch was on the camshaft instead of on the crankshaft, the normal Deere practice, resulting in the pulley running in the reverse or counter-clockwise direction.

Again because the "H" had only three forward speeds, a foot throttle pedal was provided to override the governor when used on the road. In 1940 a single-front-wheel option was offered, the "HN," and it remained in the line throughout the production life of the "H." In contrast the "HNH" and "HWH" models, with 38″ rear wheels to give extra clearance, were only built for one year, from March 1941 to January 1942. The two types of front axle for the "HWH" mentioned in an earlier chapter are illustrated.

B This "HWH," No. 29,712, has the narrower of the two front axle options

C Another view of the same tractor, "HWH" No. 29,712

A Model "H," owned by Donald Combes, at Waterloo. No. 15,250

D Earl Scott's "HN" No. 14,677, pictured in front of one of the buildings at the Waterloo, Iowa, show grounds

E The "HWH" No. 35,402 with the wider front axle, owned by Ross Johnson of Overland Park, Kansas

F A side view of "HWH" No. 35,402

G Charles Noland's electric-start "H" No. 55,656 from Adair, Iowa This picture shows the position of the battery on this "H."

1947—A Year of Great Advances, Many New Models

The year 1947 saw a number of developments in Deere's tractor business. The biggest of these was the opening of the new tractor factory in Dubuque, Iowa, built to produce a new line of small tractors to compete with the increasingly popular Ford-Ferguson.

The "M" had been on trial for some three years before production started. Fitted with a vertical 2-cylinder all-square 4″ × 4″ gasoline engine of 20 hp, it was designed to replace the "H," "L," and "LA" models. Fitted with Deere's Touch-O-Matic hydraulics, it was initially a standard 4-wheel tractor of modern design, with electric starting as standard, as was its PTO. More than 20 integral implements were offered for use with its Quik-Tatch hitching system. Lighting and a belt pulley were optional extras. The styling was the same as used on the larger row-crop tractors.

These models, the "A" and "B" series, were modernized in the same year, and fitted with electric starting and lighting as standard. This equipment had been available previously as an optional extra since the styling of these models; in fact the first styled "B" had been so equipped.

At the same time they had pressed-steel frames and were available with either all-fuel or high-compression gasoline engine. While the "A" retained the same size engine, the "B" had two new engine choices: with 4^{11}/$_{16}$″ × 5½″ replacing the earlier 4½″ × 5½″, and with speed governed at 1250 instead of 1150 rpm. Another option added was Roll-O-Matic "knee-action" front wheels for the twin-front-wheel models. This innovation halved the rise in the front end when one wheel rolled over a bump.

Ⓐ 1950 "A" with electric starting and pressed steel frames, the property of Alvie Grossman Jr., Pryor, Oklahoma

Ⓑ Shipped from the factory on November 6, 1947, and purchased new by the writer's brother-in-law, this late "A," No. 594,779, now shares exhibition honors with the oldest known Overtime —at the Lackham Museum, Chippenham, Wiltshire, England

C "B" No. 257,615, with a John Deere thresher as a backdrop

D The writer's first vintage tractor, brought from Dublin on January 10, 1959. Model "M" No. 14,021, shown recently at the Science Museum near Swindon, Wiltshire, England

A A festive view of the display floor of the John Deere Plow Company (sales branch house) on Nineteenth Street, Moline, Illinois

B Model "AN" at Waterloo with the 9.00″ × 16″ front wheel used when the 2 piece convertible pedestal was fitted. Owner is John Paulsen Jr. of Anthon, Io

C Electric-start "AW" No. 594,466 at Devizes, England, in October 1981

D Electric-start "BN" with 88 mid-mounted cultivator and 777 rear bar

E Electric-start "B" No. 240,568, owned by Mike Leitner of Edwardsville, Illinois, and the older 4-speed "B" No. 72,096, belonging to Ben Roby, Amlin, Ohio

More changes in 1947

At the same time the "G" was fitted with electric starting and lighting, but the frame was not changed nor was the flywheel enclosed. The "GM" model classification was dropped at the same time. The "D" remained unchanged except that the optional electric starting and lighting became more popular.

A "GM" No. 14,074 at Waterloo, the property of Alvie Grossman Jr., Pryor, Oklahoma

B Rear view of the same tractor

C The Model "GM" was hand started normally but the only example in the U.K., No. 20,317, has the optional electric starting and is owned by Anthony LeFanu of Bedford

D A "GW," No. 58,655, in a sale at Leesburg, Indiana, in the fall of 1984

E Author's No. 1 son posing on a Model "GN" at Etchilhampton, England, in 1951

F The only known "GW" in England, No. 28,355, in 1981

G The last "G" built, No. 64,530, owned by David Peters of Plainfield, Iowa

A One of the later "BR" tractors, No. 336,979, with electric starting and lighting. Owned by Harold Langbehn of Dysart, Iowa

B Another view of this tractor

C The view from the driver's seat on No. 336,979

D A nicely restored "D," No. 175,870, threshing at the largest U.K. vintage show at Stourpaine Bushes, Dorset, in 1987

E A very original electric-start "D," No. 175,611, at the Leesburg, Indiana, sale in 1984

F "D" streeter No. 191,661 on steel wheels with extension rims, at the Waterloo show. Owned by Charles English of Evansville, Indiana

G Beautifully restored "D" No. 177,155 in a sale near Blackpool, England. This tractor was sold in May 1987 for £4,200 ($7,500).

Deere Introduces a Diesel—1949

It was another two years before the next major developments took place. Experiments had taken place for a decade, first with a diesel version of the Model "D" and later with the MX experimental tractors, the factory designation for preproduction "R"s. When the author visited Waterloo in October 1947 rumors were rife about the new "R" diesel, the production model of the MX, but it was in June of the following year that the long-awaited tractor was announced to Deere's dealers in Winnipeg.

This 51-hp tractor was an immediate success, with its 5-speed gearbox and terrific lugging ability. It had a 2-cylinder horizontally opposed gasoline starting engine to preheat the 2-cylinder diesel $5\frac{3}{4}'' \times 8''$ 1000-rpm main engine. The starting engine was itself electrically started. Independent rear wheel brakes and an armchair seat were standard equipment; as an optional extra one could have both a fully independent PTO, operated by a separate independent clutch, and hydraulic Powr-Trol driven continuously, allowing remote operation of equipment at all times.

When tested at Nebraska, the "R" proved to be the most economical tractor tested up to that time, and remained so until another Deere model, the "70" diesel row-crop, took the honors. The "R" remained in production for six years.

B Making an interesting comparison, the only "R" imported into the British Isles originally—by Jack Olding & Co. Ltd. into Dublin in 1950 and subsequently by the writer into England in 1959—No. 4,661, is shown alongside 80 No. 8,002,858, which was also the first 80 to arrive in the U.K.

C The same Model "R" No. 4,661. At the wheel is a well-known British collector, Herbert Wilkins of Bampton, near Oxford.

A 1946 photo of "MX" tractor with single-cylinder starting engine

D E

F G

H

[D] Representing the Model "R" at the Waterloo, Iowa, Show, Ken Waits of Rushville, Indiana, showed No. 18,446.

[E] An unusual option for the Model "R"— steel wheel equipment—on a tractor at the Stockton, Kansas, show in 1986

[F] An overhead rear view of "R" No. 18,446 owned by Ken Waits of Rushville, Indiana

[G] An interesting adaptation of two "R" tractors outside the Double-R Implement Co., Estevan, Saskatchewan, Canada, in 1972

[H] An "R" at the Sully, Iowa, sale in July 1987—No. 12,668

New Regular and Orchard Models

At the same time as the "R" was introduced, the opportunity was taken to style the "AR" and "AO" models. Using the same more modern style adopted for the "R," they were completely redesigned with 6-speed gearbox, electric lighting and starting, armchair upholstered seat, Power-Trol and PTO option as on the larger tractor, and the two higher power output engines of the row-crop models. The "AO" could be purchased with elaborate fully streamlined shields to protect branches and fruit.

B A well-restored styled "AR" at the Waukee, Iowa, show on July 18, 1987

A The Kellers' very original styled "AR" No. 274,183 on steel wheels with extension rims, at home at Forest Junction, Wisconsin

C Styled "AO" No. 280,347 in fully shielded form, the property of Mike Twiss from Milton, Ontario, Canada

D Styled AR No. 277,143, shown at Waterloo by Jack Bible and David Vanaman of Newmarket, Tennessee

E Rear view of "AO" No. 280,347 showing driver protection, low seating position and Powr-Trol

F Opposite side view of the same tractor

The "M" Line Extended

The lower end of the horsepower scale was extended with the introduction of the "MT" tricycle general-purpose version, available with twin or single front wheel or wide adjustable front axle, and a new dual-type Touch-O-Matic hydraulic control to allow raising and lowering each side independently.

At the same time a new crawler tractor, the "MC," was announced. Assembled in the Lindeman factory, which the company had purchased in 1946, it replaced the "BO"-Lindeman. To an "M" skid unit supplied by Dubuque, the tracks and their controls were added at Yakima. This model represented the first Deere-designed crawler, and the beginning of a line which was to develop into the industrial division of the company. The "MI" industrial and highway model announced in 1949 further extended the "M" line.

C Well-restored "MT" No. 26,943, purchased by the writer in Lyle Dumont's sale at Sigourney, Iowa, in July 1984

A Preproduction "MT-N" No. 3 with MT-5 toolbar bedder at the Laredo, Texas, experimental farm in January 1948

D Rear view of "MT" No. 18,266 at New Alexandria, Pennsylvania, in 1978, showing belt pulley attachment

B Studio photograph of preproduction "MT-N" code XMT 01, August 1947

E Preproduction "MC" and "KB" disk harrow in December 1946 near the Lindeman factory at Yakima, Washington. Mr. Murphy of the Portland branch is at the controls and the hat of C.D. Wiman, president of Deere & Company, can be seen. He is sitting on a "BO"-Lindeman crawler.

F "MC" crawler No. 12,327 owned by Anthony LeFanu of Bedford, England, at a vintage show at Fairford, Gloucestershire, England, in 1986

G Wide-front-axle "MT-W" owned by Bob Serlet of Omaha, Nebraska

A An "MC" and dozer blade leveling ground on a new housing site

C Three-quarter rear view of the "MI" No. 10,001

B The first "MI" built, No. 10,001, in highway orange at Waterloo, Iowa

D "MI" with Charles Deere Wiman at the wheel

New Hi-Crop Tractors—1950

In 1950 the "AH" and "GH" Hi-Crop tractors were introduced for farmers with bedded crops and other specialized requirements, and these tractors represented the last lettered models to join the line, by then 20-plus strong. With a clearance of 32″ at every point under both axles and 48″ between the final-drive housings, these very high tractors were made more practical by electric starting. They were intended primarily for use in sugarcane, nursery culture, and in the growing of pineapples.

A The "GH" high-crop tractor belonging to Charles English of Evansville, Indiana, No. 47,680

B Model "GH" No. 52,934 at Lloyd Bellin's, Isanti, Minnesota, in Oct. 1986

IT'S THE FAMOUS MODEL "A" On Stilts

C Model "AH" high-crop tractor shown in a 1950 sales brochure

The Introduction of the Numbered Series—1952

The most popular tractor model in Deere's history, the "B," was finally superseded in 1952 by the 50. Similarly the 60 replaced the "A." These two new tractors, styled like the "R," had a host of new features, including duplex carburetion, hot or cold manifold, "live" powershaft through a second gear on the crankshaft ahead of the main clutch, and high-pressure Powr-Trol.

Quick-change rear-wheel tread was achieved by using a tapered sleeve on a round axle with a keyway for driving and a rack and pinion to move the heavy rear wheels. A belt-driven water-pump cooling system, and many internal improvements meant that overall another leap forward in tractor design was offered to the farmer.

Longer control levers added to the operator's comfort. From 1947 on the "B" had the single gearshift lever adopted earlier on the "G," while the "A" series followed suit in 1950.

[A] Group of numbered series tractors photographed at the Waterloo show. From right, 40 Row-Crop Utility, 50, 60 Orchard, 70 diesel, and 80 with all-steel factory cab.

A Model 50 No. 5,026,185, one of the tractors in the group on the previous page, owned by Richard Ramminger of Morrisonville, Wisconsin

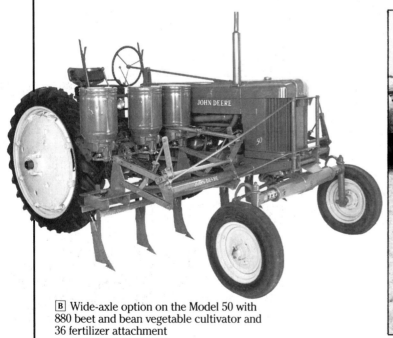

B Wide-axle option on the Model 50 with 880 beet and bean vegetable cultivator and 36 fertilizer attachment

C William Wait's 60 with downswept exhaust and special double seat used for dealer demonstrations. No. 6,057,153, from Rushville, Indiana

D 60-W with 200A beet harvester and 210 beet topper in 1956

E Single front wheel 60 tractor with an 882 cultivator in June 1955

F Jack Bible and David Vanaman showed this beautiful 60 Grove model with front grille guard at the Waterloo, Iowa, Show, No. 6,036,153

The 70 and 40 Series Announced

In 1953 the "G" and "M" series became the 70 and 40, respectively. More than 40 matched tools were available for the smaller tractors, while the 60 and 70, in addition to the all-fuel and gasoline engine options, had a third added, LP gas.

A Model 70 standard gas tractor, purchased by the writer at Lyle Dumont's sale at Sigourney, Iowa, in 1987. No. 7,023,417 has since been restored

B Model 40S standard tractor with integral 472 disk plow in December 1952

Corn Poppin' Johnny?

Merle L. Miller, Waterloo, Iowa, retired Waterloo Works engineer, recalls a unique case of tractor overheating. This was before water pumps, in the days of thermosiphon cooling which didn't tolerate much obstruction. Removing the large cast radiator cap, the troubleshooter found corn silk, husks, and a complete ear of corn in the tank on top of the radiator core. Seems the help had been heating roasting ears for lunch.

C A 40T row-crop tractor with Van Brunt integral cultivator

D Model 40T-W wide-front-axle tractor plowing in August 1955 with a 3-furrow 416 integral plow

A New Hitch

A new 800 3-point hitch was added in 1953 for the 50, 60, and 70, with a large range of new implements to make pick-up-and-go farming easier. It was the first North American introduction of the Category 2 hitch which later influenced development of a new ASAE-SAE technical standard.

New Hi-Crops

The two Hi-Crop models had changed designation with their respective series change. In addition a special 60 Hi-Crop with a fixed-tread 38″ front axle designed specifically for use with 3-row bedder equipment was added to the other front-end assemblies available.

"How It Started" Department

Water pumps on tractors made at Waterloo came in as part of the Korean War effort, 1951-53, when for a time radiator cores had to be made partly of steel instead of all copper. To compensate for the reduced cooling efficiency, the water pump was introduced. Drive was from the front end of the fan shaft.

—*Merle L. Miller, Waterloo, Iowa*

A Model 70H Hi-Crop tractor with LP-gas engine disking with a 6XH Killefer offset harrow

B Model 60 row-crop tractor fitted with the special 38″ fixed-tread front axle

C A 60H Hi-Crop in a wide bed of gladioli

D Studio photo of
70 Hi-Crop diesel tractor
in January, 1955

New Crawlers for Farm and Industrial Use—1953

When the 40C crawler replaced the "MC," a complete redesign of the track system was undertaken. A choice of 4- or 5-roller models with 83¼″ or 94⅜″ overall track length allowed a great improvement in traction, and much greater stability when using front-end equipment.

Power Steering Option Added

In 1954 Deere introduced the first built-in, factory-engineered power steering on row-crop tractors. The same year saw the addition of a 40U Utility tractor, built low for work in buildings, orchards and groves, and to give extra stability on sidehills.

In the same year the 60 Standard tractor was remodelled on the lines of the 70S with a high seating position, starting with No. 6,043,000. The higher clearance meant that the new 60 Standard was very suitable for the rice farmer as well as the grain grower.

B A popular exhibit at the Waterloo, Iowa, show was this 40 crawler fitted with a dozer blade, No. 61,902, the property of Roger Iverson, Harrisburg, South Dakota

A An example of the 40C with 5-roller tracking

C Another 40C 4-roller crawler and blade

D The earlier-type 60S standard tractor with low seat as fitted to the styled "AR" models, owned by Mr. Kepple of New Alexandria, Pennsylvania, in 1978. He is leaning on the tractor, No. 6,030,102.

E In contrast, 60S late type No. 6,046,567, seen at Waterloo, the property of Stanley Kucera of Mead, Nebraska

A Row-Crop Diesel Joins the Line—1954

An important development in 1954 was the introduction of the company's first diesel row-crop tractor, the 70 diesel. As already mentioned, when tested at Nebraska this model took the economy record from the "R" and reduced the pounds fuel per hp-hour to 0.397 when operated on the maximum belt horsepower test. The "R" figure had been 0.407.

During 1955 the 40 series was extended, with three additional models added. The 2-Row Utility model was equipped to take a 2-row drive-in front-mounted cultivator; it also had long rear axles to give a wide choice of tread. A 40 Hi-Crop tractor was added for those farmers who needed this type, and a 40 Special with 26½″ clearance compared with the Hi-Crop's 32″ offered a compromise option.

The 50 series had the LP-gas engine option added. The 801 weight-transfer hitch replaced the earlier 800 and 800A units.

B A 70 diesel tractor with wide front axle plowing with a 4-furrow 66 plow in May 1955

A Bob Waxler's 70 diesel tractor at Waterloo. No. 7,033,877 comes from Olney, Illinois with optional pre-cleaner air intake.

C Model 70 diesel standard No. 7,038,786, the property of Gerry TerHark of Freeport, Illinois, seen at the wheel

D Paul Gilsinger's 40 Row-Crop Utility at the Waterloo, Iowa, Show, No. 61,539

E A 40H Hi-Crop tractor—studio photograph in 1955

F Model 40S special high-crop tractor with 26½″ under-axle clearance, compared with the Hi-Crop's 32″

A New Large Diesel

Finally in 1955 the 51-hp "R" gave way to the new 67.6-hp 80 diesel. The new tractor had six forward speeds, "live" hydraulic Powr-Trol and "live" powershaft like the smaller members of the line, a 3-main-bearing crankshaft, and the option of dual hydraulic valves for two remote cylinders. It was also available with a factory-fitted all-steel cab.

B A well-restored 80, No. 8,002,970, at the Boswell Rotary Club's vintage show in Indiana in the fall of 1984

A A restored 80, No. 8,000,580, at the 1987 Sigourney, Iowa, sale

C A tractor with all-steel factory-fitted cab, 80 No. 8,002,096 at Waterloo, Iowa, in July 1987

D A Danish farm student plows with one of Don Dufner's Model 80 tractors and 5-furrow Deere plow. The whole of Don's farm at Buxton, North Dakota, is farmed with John Deere 2-cylinder tractors from his very large collection.

E The first 80 to be imported to the U.K. was No. 8,002,858 with oversize 23.1″ × 26″ rear tires. Now the property of Henry Roskilly, it is pictured here at Tavistock, Devon, England.

The Demand for More Powerful Tractors Met—1956

The 420 series was initially announced late in 1955 and continued the 10 models and the all-green color scheme of the 40 line. With their increase in power from the 40's 25.2 belt hp to 29.2, a requirement for a lower-cost, lower-powered tractor was evident, so the 40 was effectively continued as the new 320, available in both standard and utility form. No new Nebraska test was deemed necessary.

[B] A 320U utility tractor, No. 325,306, at the 1987 Waterloo show

[A] A 420 Row-Crop Utility tractor plowing with a 3-furrow 416 integral plow

[C] A studio photograph of a 420S standard tractor

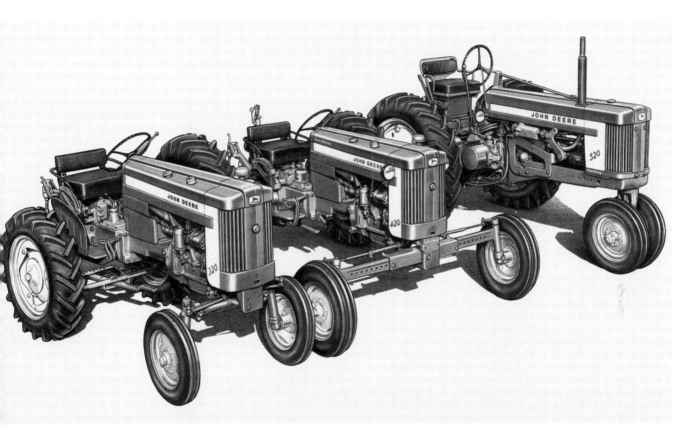

D Group studio photograph of Model 320S, 420T with wide front axle, and 520 row-crop tractors

E A 420 Hi-Crop with clearance of 32″ under the axles and 48″ between the rear axle housings

F A 420 Hi-Crop in the original all-green color scheme at Arthur Bright's farm, Le Grand, California, in 1982. No. 91,804

New Options for the 420

The 420 models added a number of new optional features when announced. A 5-speed transmission, continuous-running PTO, dual Touch-O-Matic for the standard model, and a 3-point hitch for the crawlers were some examples of the effort to widen usefulness of this series. Later an LP-gas engine option was added.

B Studio photograph of 420C 4-roller crawler

A A 420 Hi-Crop tractor cultivating in February 1957 in the type of crop for which it was designed

C Plowing on the contour, a 420C 5-roller crawler and 66 plow

D 420C No. 58,441 with "MC" No. 9,825 and "BO"-Lindeman No. 335,469 on the Deere & Co. Administrative Center's display floor on August 3, 1987

E Disking with a 420T LP-gas tractor and mounted "C" disk in August 1957

F The same tractor and disk showing the 3-point hookup

The Other 20 Series Announced

Announced in the summer of 1956 for the 1957 season the rest of the "20 series" tractors were introduced, giving an average power increase of 20%, and achieved a new standard of performance. The exceptions to this were the 320, mentioned above, and the replacement for the 80, only recently announced, which became the 820.

In addition to their more powerful engines they had Custom Powr-Trol, completely independent of the transmission clutch and PTO, giving the possibility of two rear hydraulic outlets with an emergency disconnect system.

B 520-N single-front-wheel tractor with 22 cotton picker mounted on it

A A 520 belonging to Don Keck of Menomonie, Wisconsin

C 620 gas model drilling with DR 16×8 grain drill controlled from the tractor's Powr-Trol

D 620-N single-front-wheel version with 52 Culta-Carrier

E 620-W with 30 Double Level-Bed potato digger with stone picker attachment

F 620 Standard No. 6,209,812 at Waterloo, owned by John Kahrhoff of Carlyle, Illinois

G Rear view of the same tractor showing its 3-point linkage

Customer Comfort Considered

The provision of a front-mounted rockshaft meant that, with the dual-remote-cylinder option, front-mounted integral equipment could be operated independently to overcome uneven penetration. A new Universal Category 2 3-point hitch was the first draft-responsive hitch available, while a pedal-operated independent powershaft and a new Float-Ride seat added to the farmer appeal of these new tractors.

From the 520 on, models had power steering as standard, and the whole line was attractively finished now in a new green and yellow eye-catching color scheme.

B A 720 gas-engined tractor plowing with 5-furrow 666H plow

A Studio photograph of 720 diesel cutaway model built for demonstration purposes

C Model 720-W No. 7,205,842, beautifully restored by Marvin Schaffer of Northfield, Minnesota

D Another 720-W in the studio in October 1956

E An April 1957 studio photo of the 720 Hi-Crop diesel tractor

F Advertising photograph of the new 620 and 720 Hi-Crop tractors fitted with power steering as standard

The Last 2-Cylinder Orchard Model

Late in the year the 60 Grove model became the 620 Grove and so remained until the end of the 2-cylinder era. Tailor-made for operating in groves and orchards, it incorporated full shielding including the engine, a choice of lug-type or low-profile tires, and the same three fuel choices and other options of the 620 series.

The Largest 2-Cylinder Tractor

Initially fitted with the same engine as the 80, the 820 from No. 8,203,100 on was fitted with an engine having increased horsepower, from 67.6 to 75.6, and again it achieved excellent fuel economy figures at Nebraska, only the 720 diesel being marginally better—0.388 and 0.383 lbs. fuel per hp-hour, respectively.

A Research and Engineering Center Opened

No less important was the 1956 introduction, for the first time in the tractor industry's history, of a research and engineering center devoted exclusively to the design and testing of tractors. This new Deere operation was based in the western outskirts of Waterloo, and was the home of the new series of multi-cylinder tractors being developed. But first the year saw the final flourish in the history of the 2-cylinder tractor as it had been known and loved for over 40 years.

A The longest lasting of the "20 series" models, the 620 Grove tractor, which continued in production to the end of the 2-cylinder era

B The same 620 Grove tractor disking in an orange grove with GR 716 disk harrow

C Model 820 diesel tractor at the Waterloo, Iowa, show, owned by John Ripberger of Tipton, Indiana. No. 8,200,291

D 820 No. 8,204,484 at a Leesburg, Indiana, sale on August 18, 1984

SHUT OUT *Wind, Cold, Snow, and Rain . .*

In hot summer months, swing open the windows and pull the curtain back on your John Deere Tractor Cab. You'll enjoy cool cross ventilation without the heat or glare of direct sunlight.

E From a 1957 sales brochure, details of the all-steel cab available for the larger standard tractors

The Last of the Famous 2-Cylinder Line, the "30 Series"—1958

Without any alteration in the engine size, the new "30 series" announced in 1958 offered increased operator comfort and a new and very attractive styling. A sloping automobile-style steering wheel, easy-to-read instrument panel, a new-style muffler and optional muffler cover, and fenders with dual headlamps on the row-crop models were just some of the standard features of this new line.

More-convenient hydraulic control levers could be mounted at either side of the seat. Added to the previous optional extras were power-adjusted rear wheels, Weather-Brakes and Weather-Brake cabs. A number of other extras were all intended to improve the lot of the operator.

A "30" series group of tractors at the Waterloo show. Lower left front, 330 Standard No. 331,006 430 Row-Crop No. 155,528. Upper left rear, 530 No. 5,300,447, 630 LP Standard No. 6,313,925, 730 diesel electric start No. 7,319,492 and 830 No. 8,304,611

A One of Lowell Kroneman's seven "30 series" tractors at the Waterloo, Iowa, show, his 330S No. 331,006

B A valuable load of three restored 330s ready for delivery from a Kentucky collector

C 430T studio portrait in July 1958

D This 330S at the Waterloo, Iowa, show was No. 330,228, belonging to the Polk Brothers of Leesburg, Indiana

E Gib Mouser's 430U Utility model No. 155,447 from Blackduck, Minnesota

More Extras for 430 Series

The 430 series had more options than their predecessors including a direction reverser, power steering, and a Float-Ride seat. An auxiliary hydraulic system for the operation of a remote cylinder gave the series the same flexibility as the larger models. For orchard work a downswept exhaust continued as an option.

B 430C cutaway from 1958 sales literature

A 430T No. 158,244, one of the Kentucky collection

C 430C 5-roller crawler plowing in May 1958 with a 4-furrow 555H plow

D 430HC Hi-Crop tractor fitted for LP-gas fuel. Owned by Verlan Heberer of Belleville, Illinois

E Another view of the same tractor, No. 140,822, now fitted with a sunshade

F 430T-N single-front-wheel model with experimental one-row corn picker

A Small Diesel

The last 2-cylinder model to be added to the line in 1959 was a GM-diesel-engine version of the 430 general-purpose tractor, the 435. With a $3\frac{7}{8}'' \times 4\frac{1}{2}''$ engine governed at 1850 rpm, it had the distinction of being the first tractor tested at Nebraska after the introduction of the new standards for 540/1000-rpm power take-off shafts had been introduced. It produced 32.91 hp at the PTO, compared with 29.72 at the belt on the 430 model.

B Lowell Kroneman's 435 with GM diesel engine. No. 436,683

A A 430 Row-Crop Utility tractor disking in May 1958 with a 24-disk model KBL integral disk harrow

C Right-side detail of the GM diesel engine on another 435, No. 438,558

D A 435 at work mowing and conditioning a field of alfalfa on the Dufner farm at Buxton, North Dakota

A Lowell Kroneman's
530 No. 5,300,447

B A 530 with LP-gas engine
at the Waterloo, Iowa, show,
owned by Paul Franken of
Clarksville, Iowa

C 630 No. 6,300,784, another of Lowell Kroneman's collection

D A 630 tractor equipped for LP gas, with a 555H 4-furrow plow in May 1959

A Comprehensive Choice

The final 2-cylinder line was complete—three standard fixed-tread tractors, the 630 4-plow, 730 5-plow and the mighty 830 6-plow for the grain and rice growers, fourteen general-purpose models from the 1-2 plow 330 to the 5-plow 730 for the bean and corn "growers," two 430 crawlers for use where wheel-type tractors could not go, three high-crop models and three low-built tractors for specialist applications.

In addition to more than thirty basic agricultural tractors, a lengthening yellow line of industrial models was offered for construction work and highway maintenance, but more of this in the next book.

[B] A 630S standard gas tractor, No. 6,310,957, in a sale in 1984 at Leesburg, Indiana

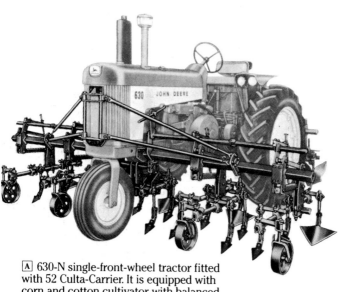

[A] 630-N single-front-wheel tractor fitted with 52 Culta-Carrier. It is equipped with corn and cotton cultivator with balanced draft to ensure precision work without side-drifting.

[C] Danny Witter showed this 630S LP standard No. 6,313,925 at the Waterloo show. He hailed from Mercersburg, Pennsylvania.

D 630-W baling hay with a 214T pickup baler

E The 630 Hi-Crop was available with gas, all-fuel or LP-gas engine. Here it's equipped with a special Hi-Crop hitch in addition to the standard tractor's power steering, Powr-Trol, independent PTO and Float-Ride seat.

A A 730-W wide-front-axle model with a 5-furrow 666H plow in March 1959

B Fixed-front-axle 730S with RW wheel-carried disk harrow, working during the fall of 1958

C Cutaway show model of the 730 diesel tractor

D Rear view of this exhibit

E A 730 Standard gas tractor with adjustable front axle, No. 7,326,749

F Group of three 730 diesel tractors with suggested JD medallions—not adopted—and consisting of a 730-W with wide front axle option, a standard twin-wheel row-crop, and a Hi-Crop with special 38″ fixed-tread front axle

G The Kentucky collection's 730S diesel model, missing its diesel decal, and with adjustable front axle. No. 7,310,709

H 730 Standard LP-gas tractor with adjustable front axle at Dennis Polk's Leesburg, Indiana, sale in 1984. No. 7,328,638

The End of an Era

And so we all went to Marshalltown, Iowa—
farmers, dealers and overseas guests—to admire
the latest tractors, combines, balers and all the
other implements which, in 1959, made up The
Long Green Line. We little realized that the days
of the 2-cylinder tractor, which had lasted for
over 40 years, were numbered. Even the presence
of an enormous 4-wheel-drive tractor pulling a
fully mounted 8-furrow plow did not provide
a clue. It merely meant that a number of marshals
had to be posted at the furrow ends to prevent
anyone from getting hurt by the enormous swing
of the plow…

B The ultimate model 2-cylinder tractor, the 830 diesel with
V-4 auxiliary engine starting. This top-of-the-line model
completed the Kroneman collection at Waterloo. No. 8,302,830

A Plowing five furrows on the beautiful dark loam soil on Don
Dufner's farm. One of his sons is at the wheel of the 830, in 1986.

C Belt-pulley-side view of the same tractor

D An interesting contrast with the agricultural tractors was this industrial-version 830I No. 8,303,801 from Charles Smith of Des Moines, Iowa, later owned by the Bellins of Isanti

E A beautiful 840 Industrial tractor, No. 8,400,387, with Hancock self-loading scraper. These were the last 2-cylinder tractors built for the home market through 1964. Owners were Fischer Farms of Watseka, Illinois

Tillage Review

Plows—the Beginning

At this point in our story we must return to the beginning and the plow. The first crudely shaped wooden frame, with its converted sawblade bottom, was eventually replaced by a plow with more shapely handles and beam. It was many years before these parts were replaced with steel.

The simple bottom soon gained a replaceable cutting edge or share, but it was another 30 years before the first "ride-on" or sulky plow was introduced. Gilpin Moore was the best-known designer of these new plows, which remained in production, largely unchanged, for the next 60 years.

Single- and 2-furrow versions were available. By ganging the latter into groups, giant engine plows up to 12- and 14-furrow size were developed before the end of the nineteenth century for use with steam traction engines and later with the large early gas tractors.

B 1838 plow and anvil in Deere & Company display-floor mural

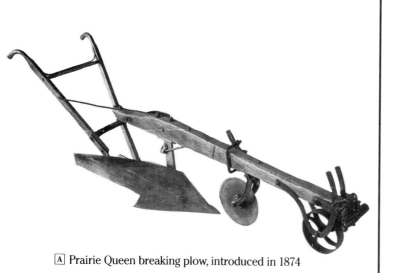

A Prairie Queen breaking plow, introduced in 1874

His First Lesson

The farmer of today proudly teaches his son what his own father taught him— to use a John Deere Plow.

C "His first lesson"

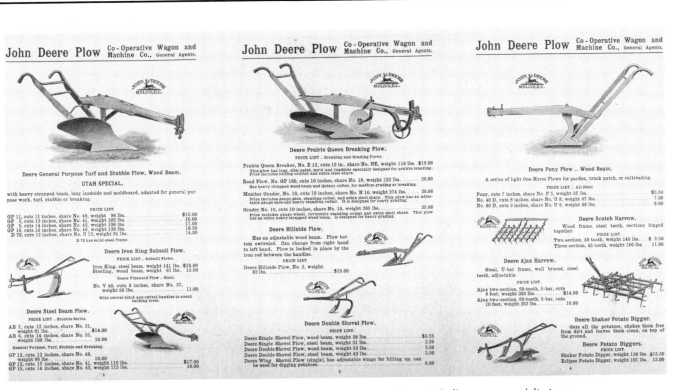

D A 1901 advertisement for various wooden- and steel-beam walking plows, including many specialty types

E Plowing turf with a steel-beam walking plow and two horses

A Kid one-furrow sulky plow, shown in a Kansas City branch catalog

B The Gazelle gang plow of 1894

C Engine gang plow fitted with eight stubble bottoms, from a Dallas catalog

[D] A restored 2-furrow sulky plow in a sale at Sigourney, Iowa, on July 28, 1984

[E] Two-way sulky plow, from 1910 St. Louis branch catalog

[F] Demonstration of a 10-furrow engine gang pulled by a kerosene-burning Russell Giant 30-60 tractor made in Massillon, Ohio

Disk and 2-Way Plows Introduced

Both 2-way and disk plows were introduced, initially for horses, but later for both engine and tractor use. Announced during World War I, the Pony Nos. 5 and 6 models soon became the most popular tractor plows, and so remained until replaced by the 55 and 66 Truss-Frame models introduced just before World War II.

To meet the needs of the smaller tractor user, the 2-furrow No. 4 series was offered, and a special Model 40 was announced in the early twenties for use with the Fordson, which was increasingly popular at that time.

All these models were updated in 1941 with the 44, 55, 66 and 77 Truss-Frame plows already mentioned, and these gave way in turn to the 444, 555, 666 and 777 series before the end of the 2-cylinder tractor era.

With the introduction of the 3-point hitch as regular equipment on row-crop tractors and frequently an option on standard types, integral plows became increasingly in demand.

The 3-furrow 416 shown on a 435 tractor at the Marshalltown John Deere demonstration in 1959 is a good example of these units.

A A 1931 picture of the 2-way No. 1 plow supplied for the Model "GP"

B Some 27 years later the 3-furrow 2-way 825 plow mounted on a 730 tractor makes an interesting comparison with the picture above

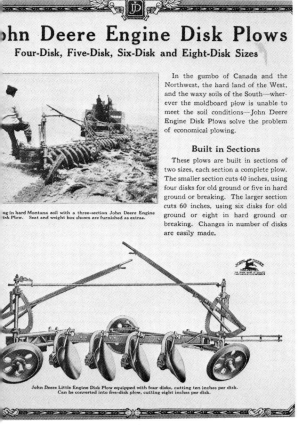

ohn Deere Engine Disk Plows
Four-Disk, Five-Disk, Six-Disk and Eight-Disk Sizes

In the gumbo of Canada and the Northwest, the hard land of the West, and the waxy soils of the South—wherever the moldboard plow is unable to meet the soil conditions—John Deere Engine Disk Plows solve the problem of economical plowing.

Built in Sections

These plows are built in sections of two sizes, each section a complete plow. The smaller section cuts 40 inches, using four disks for old ground or five in hard ground or breaking. The larger section cuts 60 inches, using six disks for old ground or eight in hard ground or breaking. Changes in number of disks are easily made.

ng in hard Montana soil with a three-section John Deere Engine isk Plow. Seat and weight box shown are furnished as extras.

John Deere Little Engine Disk Plow equipped with four disks, cutting ten inches per disk. Can be converted into five-disk plow, cutting eight inches per disk.

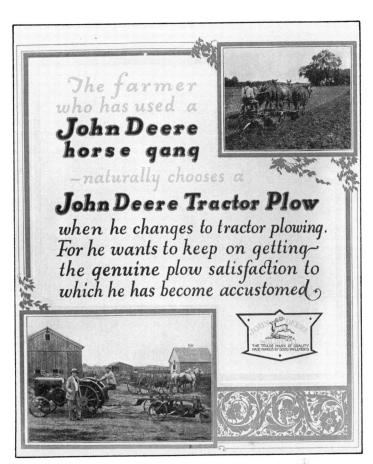

The farmer who has used a **John Deere horse gang** *—naturally chooses a* **John Deere Tractor Plow** *when he changes to tractor plowing. For he wants to keep on getting the genuine plow satisfaction to which he has become accustomed*

THE TRADE MARK OF QUALITY
MADE FAMOUS BY GOOD IMPLEMENTS

C A 1926 Plow Works advertisement for their Engine disk plows

D Another 1926 Plow Works advertisement, this time for the long-lived Nos. 5 and 6 tractor plows

E Waterloo Boy "N" plowing with a 3-furrow No. 5 plow in June 1920

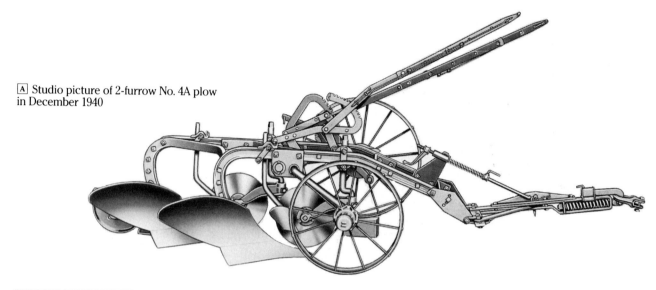

A Studio picture of 2-furrow No. 4A plow
in December 1940

C A 6-speed Model "B" tractor with 2-furrow plow on rubber
tires at the same show

B Single-furrow 51 plow at the Waterloo
celebrations in July 1987

D The author enjoying himself in 1959 at the John Deere demonstration at Marshalltown, Iowa, plowing with an 830 tractor and 6-furrow 777H plow

E Bob Lovett, formerly of the Deere Export Department, at the wheel of a new 435 diesel tractor with 3-furrow 416 integral plow at the Marshalltown event

F Another 830 tractor pulling a 6-furrow 777H plow

The Plow Line Expands

Many variations of the basic plow were offered including prairie breakers, brush breakers, grading and mole plows, chisel plows for heavy cultivation, and special types for digging potatoes. With over 180 types of plows on offer at the beginning of the century and an equally comprehensive coverage in succeeding years, farmers' requirements were fully satisfied.

Plow bottoms were offered in several different types. Stubble and general-purpose were the most popular, the latter available with moldboard extension to assist in turning tame sod, while specialty bottoms for breaker plows and either slatted or rod for certain soil types fitted all plow models.

B John Deere Killefer 32-06 mole plow at the Waterloo show, July 1987

A A Model 12 Brush Breaker plow cutting a full 16 inches

C A 15 subsoiler at the same show

D Model 830 tractor chisel-plowing with a 656H tool carrier in May 1958

E A 900 tool carrier with coil-spring tines behind a 620 tractor with Powr-Trol

F A 15 subsoiler at work behind a 620 tractor

Cultivators

Cultivators were the first departure by John Deere from plow construction, and by the beginning of the twentieth century the company was the largest producer in the United States, with 20% of the total market sales.

The early Hawkeye cultivator of 1858 gave way to the Peerless in 1876, and later to spring-tine machines. After its introduction the "CC" became very popular and stayed in the line into the '40s, only being replaced by its updated version the "CC-A."

A An 1874 advertisement for the Advance cultivator and the plows which took first prize at the Vienna Exposition in 1873

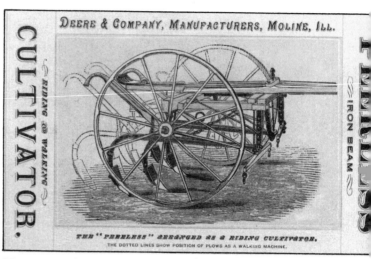

B Peerless riding (as shown) and walking iron-beam cultivator, 1875

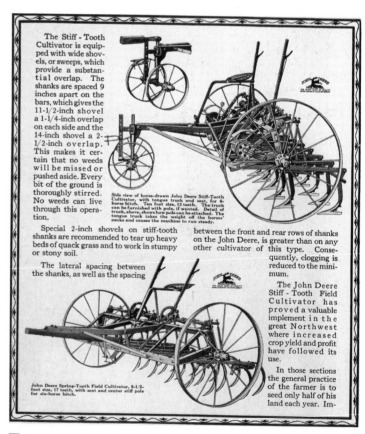

The Stiff-Tooth Cultivator is equipped with wide shovels, or sweeps, which provide a substantial overlap. The shanks are spaced 9 inches apart on the bars, which gives the 11-1/2-inch shovel a 1-1/4-inch overlap on each side and the 14-inch shovel a 2-1/2-inch overlap. This makes it certain that no weeds will be missed or pushed aside. Every bit of the ground is thoroughly stirred. No weeds can live through this operation.

Special 2-inch shovels on stiff-tooth shanks are recommended to tear up heavy beds of quack grass and to work in stumpy or stony soil.

The lateral spacing between the shanks, as well as the spacing between the front and rear rows of shanks on the John Deere, is greater than on any other cultivator of this type. Consequently, clogging is reduced to the minimum.

The John Deere Stiff-Tooth Field Cultivator has proved a valuable implement in the great Northwest where increased crop yield and profit have followed its use.

In those sections the general practice of the farmer is to seed only half of his land each year. Im-

C Two views, one of a horse-drawn stiff 13-tooth 10′ size cultivator and the other a spring-tooth field cultivator with 17 teeth, 8½′ size and with 6-horse hitch

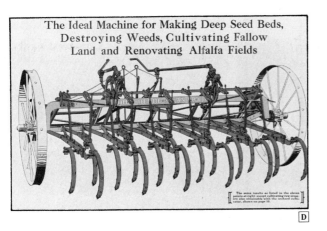

The Ideal Machine for Making Deep Seed Beds,
Destroying Weeds, Cultivating Fallow
Land and Renovating Alfalfa Fields

D

D A 1933 advertisement for the famous Van Brunt "CC"
general-purpose field cultivator

E A Model "A" tractor in charge of a rubber-tired "CC" orchard
model cultivator with trailing harrows in September 1936

F Another "A" tractor with a 23-tine "CC" cultivator
in May 1937

G Nearly 20 years later a 29-tine "CC" does a good job behind
a wide-front-axle 70 diesel tractor

E F

G

Disk Harrows

Early horse-drawn disk harrows were featured first in the 1890s. Marketed originally by Deere & Mansur, the Model "B" disks were famous for many years, and were still in the 1942 sales catalogs.

Tractor-drawn Pony disks up to 10′ wide preceded a number of different models offered in the '20s and '30s, including the single-action "S" type up to 21′ wide, the standard-weight "J" series, the heavier-duty "K" model, and specialized offset disks for orchard and grove work.

Over the years the width of disk harrows increased with the increase in tractor power. The addition of remote hydraulic control also meant that disks could be angled from the tractor seat, and that wheel-carried harrows could be raised and lowered in the same way. If sufficient width was not obtainable in one unit, squadron hitches were used to combine two or more sets.

As with plows so with the other tillage equipment: the advent of rear linkage on tractors meant the introduction of integral units to match. Disk harrows were no exception, and the "KBL" double-action and the offset M-404 for the smaller tractors introduced this trend. The 414 series replaced the latter in the '50s, and the "KBY" and 416 were added for the larger Waterloo tractors.

[B] The well-known Model "B" single-action 16-disk harrow, pulled by four horses in August 1918

[A] Pony double-action disk harrows with yielding lock to prevent skidding of the rear section when turning, pulled by a Waterloo boy "N" tractor in 1923

[C] The famous Model "S" 21′ single-action disk harrow, shown behind a styled model "D" in April 1940, was still in the line in the mid-fifties

D An October 1947 picture of the standard-weight "JB" double-action disk harrow behind a late-type Model "A"

E A 28-disk "JB" double-action harrow and late "B" tractor prepared for the John Deere Moline centennial celebrations

F A 40U utility tractor with one orchard fender and "SH3" offset disk harrow built in the Killefer Works. The "H" series were the first disk harrows to be equipped with wheels for raising and lowering.

[A] A 720 standard diesel tractor equipped with 3-point linkage and Powr-Trol, disking in October 1956 with a Killefer heavy-duty offset harrow fitted with cutout blades on the front gang

[B] Studio photo in November 1952 of the integral "KBL" disk harrow mounted on a 40T tractor

[C] Another version of the "H" series disks, the "6XH," also built by Killefer, behind a 70 Hi-Crop tractor equipped for LP-gas, in April 1955

D A 720 makes light work, in August 1957, of pulling a dual-wheel "FW" double-action disk harrow, which could be reduced in width for transport by folding in the outer frames

E In May 1958, a 730 is in charge of a similar rig

F To obtain a wider cut, two 12′ 1400 series Killefer-built disks are shown with a squadron hitch behind a Caterpillar track-type Tractor

Disk Tillers and Integral Cultivators

Disk tillers were popular in the wheat-growing regions, with seeders attached for a once-over operation in many areas. By the end of the fifties they had grown to 20′ in width and were available in many types, the heavier-duty models being built in the Killefer Works.

With the introduction of mechanical and subsequently hydraulic lift on tractors, integral cultivation equipment, toolbars and carriers became very popular for row-crop work, starting with 2-row and expanding to 8-row during the period covered.

B A series 1100H Surflex tiller behind a standard 70 diesel tractor

A This 1200 series Surflex tiller is equipped with the optional rubber-tired transport wheel and is pulled by an 820

C A side view of the outfit in picture A, in July 1957

D A June 1957 picture of another 820 with a hydraulically operated Surflex tiller without the optional transport wheel

E This 35-disk 1200 Surflex tiller cutting 20′ was the largest model built in 1957. The motive power is again an 820.

F In May 1958, a 20′ 1200 Surflex tiller is matched here with an 830

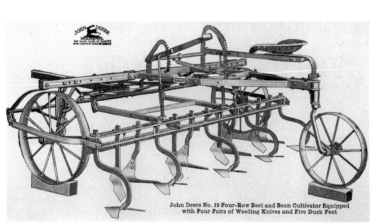

John Deere No. 19 Four-Row Beet and Bean Cultivator Equipped
with Four Pairs of Weeding Knives and Five Duck Feet

A

B

A The 4-row 19 beet and bean cultivator, in 1923 catalog

B The 1958 answer—a 6-row 60 cultivator mounted on a 730 fitted with Roll-O-Matic front wheels

C Two views of 2-row GP-422 cultivator fitted to a Series "P" tractor, 1930

GP-422 Two-Row Cultivator—Spring Trip.
Row spacing 34″, 36″, 38″, 40″ and 42″, adjustable on front pipe.   Front rigs, 8 shovels, spring trip.   Front rigs have compression springs for depth control.

wide, 3-1/4″, and only a short distance apart, 13-1/2″. There is no springing of the crankshaft under the heaviest loads.

The simple, heavy-duty engine in this General Purpose Wide-Tread Tractor meets all the requirements of balance, flexibility and smooth, efficient operation—this has been thoroughly proved.

Rear View GP-422 Two-Row Cultivator. Notice how this outfit
forms a single unit with the tractor.

Full Force-Feed Pressure Lubrication System

Oil is forced under pressure to the main bearings, connecting rod bearings through the connecting rods to the piston pins and through an additional oil pipe to the governor housing for the lubrication of all governor parts. This oil pressure is sufficient to prevent metal-to-metal contact of the crankshaft with main or connecting rod bearings, and the piston pin with connecting rod.

Oil thrown from the connecting rod bearings thoroughly lubricates all other parts of the engine.

An oil indicator in plain sight of the operator shows him that the oil is circulating. Both oil pump and oil strainer are accessible from the outside of tractor.

This efficient oiling system is an important factor in the continued full power supplied by the rugged engine with minimum wear and adjustments.

Simple and Effective Thermo-Syphon Cooling System

The engine is water-cooled by the simple thermo-siphon principle, using a tubular radiator. This provides heat control in the simplest, most effective way. This system does away with fan belt and water pump.

Mounting the radiator high above the cylinder

C

D Overhead view of 4-row integral cultivator with fertilizer attachment mounted on a Model "B" tractor in February 1938

E In contrast, a 480 Quik-Tatch cultivator is shown on a 40T tricycle tractor in 1954

D

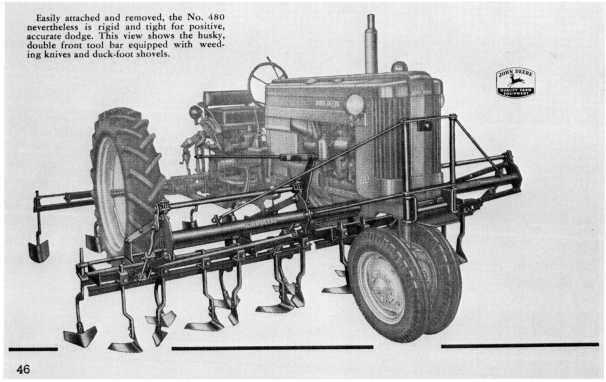

Easily attached and removed, the No. 480 nevertheless is rigid and tight for positive, accurate dodge. This view shows the husky, double front tool bar equipped with weeding knives and duck-foot shovels.

46

E

Stalk Cutters, Rod Weeders and Harrows

Other cultivation equipment popular in different areas included stalk cutters to deal with the residue of the corn and cotton crops. The Model 14 rolling stalk cutter gave way to tractor-drawn 2- and 4-row models, and after early and somewhat crude attempts at a power-driven version the No. 5 integral PTO-driven type was in the line for many years. Rod weeders were used where summer fallowing was the practice. All types of spike and spring-tooth harrows, roller harrows, and rotary hoes were some of the other tillage equipment available. To cover all the different models offered over the first 60 years of this century would be impossible in a work of this size.

B Introduced in the mid-thirties, the all-steel 2-row stalk cutter is seen cutting cotton stalks. It is pulled by a Model "B."

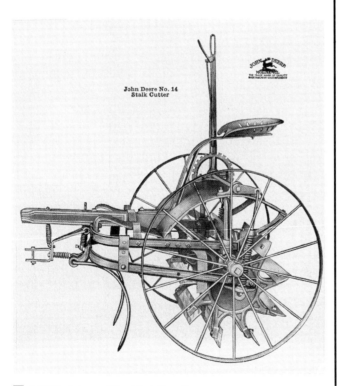

A A 1927 picture of the 14 stalk cutter

C Four-row power-driven stalk cutter designed for the "GPWT" tractor

D June 1938—two rolling stalk cutters with special hitch cover 140″ to take advantage of the power of the Model "A"

E A later development, the No. 5 integral 2-row stalk cutter with PTO drive from the 60 tractor

A Model "B" roller harrow behind a 620 LP-gas tractor in July 1957

B A gang of four No. 30 spike-tooth harrows with integral hitch on a 530. The diamond-shaped teeth are welded to the frames.

C A 12′ roller harrow in May 1958, pulled by a 730

D Covering a 48′ sweep with a multiple hitch and eight spike-tooth harrow sections behind an 830 in May 1958

E Model 314A rotary hoe does a good job behind a wide-front-axle 620

F A 530 with power-adjusted rear wheels is sufficient power for this 412A rotary hoe

Planting Review

Planters

Planting tools were another diverse subject. One-horse planters gave way eventually to the early corn planter for horse or tractor operation. Marketed through the Deere & Mansur organization, these planters quickly gained a reputation as the best on the market.

The famous No. 9 was developed into the best-known of all planters, the 999, using the wire check-row system. Still offered in 1945, this world-famous planter gained its reputation because of its exceptional accuracy. The 2-row 290 and 4-row 490 drawn models which followed were augmented by many corn and cotton planters available as attachments for integral cultivators.

Listers, middlebreakers and bedders were offered in both integral and drawn types following the earlier horse-drawn models. Sizes to suit both the smaller and larger tractors in the line meant that in 1951 there were no less than 23 different series in the catalog.

A A very early rotary-drop corn planter, one in Deere's historical collection, pictured on the Deere & Company display floor in 1987

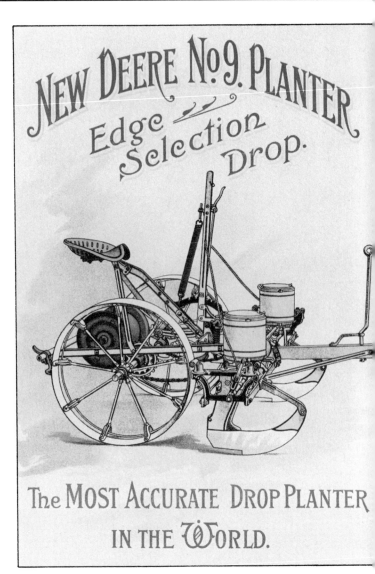

B The No. 9 edge-drop planter quickly gained a reputation as one of the most accurate planters in the world

238

MORE AND BETTER CORN

*John Deere
No. 999
Planter*

**With Pea Planting
Attachment**

Equipped with Safety
Fertilizer Attachment
and Pea Attachment.

C Replacement for the No. 9 was the 999, which became world famous and remained in the line until after World War II. It would handle corn, beans, peas, sorghum, feterita, beet seed, shelled peanuts and other seed without having to change the cutoff.

D The 999 with fertilizer and pea attachment

JOHN DEERE
MOLINE, ILL.
THE TRADE MARK OF QUALITY
MADE FAMOUS BY GOOD IMPLEMENTS

JOHN DEERE
NO.645
TWO-ROW
LISTER

Field-Proved Advantages That You Want

PAYS FOR ITSELF FROM THE *EXTRA BUSHELS* DUE TO BETTER PLANTING

E A 1927 advertisement for the 2-row 645 lister with 999 seeders and adjustable for 36″, 38″, 40″ and 42″ rows

F By comparison, a 50 series integral 2-row lister is seen in August 1952, mounted on the new 40T tricycle tractor

239

New Planters Added

The fifties saw the introduction of the 4-row 494 and 6-row 694 drawn units; the former could be "doubled" to give 8-row capability with a special drawbar which also allowed tight cornering. Integral Model 2-row 246 and 4-row 446 planters also had the famous John Deere Natural-Drop seed plates, and were for use as drill or hill-drop planters, while if you wished to plant both corn and cotton the 247 and 447 were the answer.

In semi-arid areas the Nos. 6 and 7 lister planters were preferred, while the 480 and 680 rear-mounted models were for use with tractors fitted with Powr-Trol. One other option was to equip your integral cultivator with a planting attachment. These one-, 2- and 4-row drill planters could handle corn, cotton, beans, peas, peanuts and many other crops.

B The 694 planter could drill, hill-drop or check-plant six rows at speeds up to 5 mph. One is seen here in May 1957 with a 720 diesel tractor.

C The 684 was an integral corn and cotton planter, fast on- and-off, with each planting unit individually mounted. A 630 LP- gas tractor is seeding corn in April 1959.

A A 4-row GPA-402 planter on a Model "A" tractor in May 1937. These planters were designed primarily for planting on beds but could be used in furrows as well.

D A very early No. 7 beet and bean planter

E By contrast a late electric-start "G" pulls a rubber-tired 66 beet and bean planter in June 1948

Specialty Planters

Specialty beet and bean drills added to the types on offer. The Lindeman Plantrol vegetable planter and the Hoover one- and 2-row (eventually 4-row) potato planter—the planter with the 12-arm picker wheel—gave a full line of row-crop planting equipment.

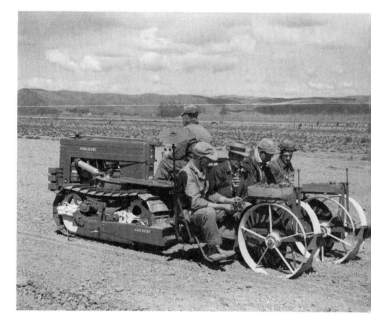

A Appropriately, an "MC" crawler built in the Lindeman factory pulls a Lindeman Plantrol transplanter mounted on an MC1000 tool carrier

B A 2-row Hoover potato planter equipped with quick-turn hitch is pulled by a "GPWT" Series P tractor in May 1931

C Six horses cope with this 20-row double-disk Van Brunt grain drill. Taken from the front page of a 1926 catalog.

Drills

The famous Van Brunt drill line, which had achieved a 50-year reputation for quality seeders, was acquired in 1911. The horse-drawn single- and double-disk drills were adapted for tractors first, and later press drills were added. As early as 1919 a 20-row 4″-spaced grass drill was cataloged, the forerunner of the "OO" which remained in the line until after World War II. The first direct drill for grass, the "GL," was introduced in 1954 and updated in 1956 as the "GL-A."

B In February 1919 a Waterloo Boy tractor has charge of an 18-row Van Brunt grain drill

A In June 1936 this "OO" grass seed drill is still pulled by two horses

C Another 1919 picture showing a horse-drawn Van Brunt low-down press drill

John Deere-Van Brunt Double Disk Drill

Sizes—Double Disks: 6-inch Feeds, 12, 14, 16, 20, 22, 24 Disks; 7-inch Feeds, 10, 11, 12, 14, 18 Disks; 8-inch Feeds, 8, 12 and 16 Disks.

D A 20-row Van Brunt double-disk drill, also available in 12- to 24-disk sizes with 6″ spacing, 10- to 18-disk with 7″, and 8-, 12- or 16-disk with 8″ spacing

E A 520 tractor with a "GL" grassland drill for direct seeding into poor pastures

F "GL-A" drill advertising pages from sales literature

Grain Drills

Grain drills included the "EE" and "EN" series with 6″ or 7″ row spacing, "SS" series deep-furrow models featuring double-row feeds, and "PD" Plow Press drills for use with 2-, 3- or 4-bottom plows. Other press drills were the "LL" for semiarid conditions, superseded by the "LL-A" and then the "LL-B."

A A "B" tractor on front rubber tires and rear steel wheels drills with a Model "EE" grain drill

B The "B" drill replaced the "EE" in the line and is shown here in May 1958 in 16×7 form with liquid fertilizer attachment. The tractor is a 530 gas model.

C In 1958 the "LL-A" press grain drill replaced the "LL," which had been in the line for 20 years. Two of these new drills are seen here behind an 830 tractor, fitted with liquid fertilizer attachment and special multiple hitch.

D The "LZ" series of lister grain drills were designed to drill through mulched topsoil where moisture conservation is important. The tractor pulling this 20-row setup is a Model "R."

Combined Drills

The requirement to seed and fertilize at the same time was met by the early combined drill. This became the wooden-box Model "F," followed by the all-steel "FF" and finally the low-wheeled models "FB," "FB-A" and "FB-B."

Drills had come a long way, in diverse forms, since the introduction of the first Van Brunt shoe drill in 1890, the single-disk type in 1900, and the adjustable-gate force feed which handled seeds of varying shapes and sizes.

A This one-horse Model "X" 5-disk combination grain and fertilizer drill is shown on the Deere & Company display floor in 1987.

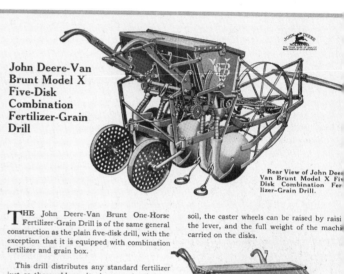

John Deere-Van Brunt Model X Five-Disk Combination Fertilizer-Grain Drill

Rear View of John Deere Van Brunt Model X Five Disk Combination Fertilizer-Grain Drill.

THE John Deere-Van Brunt One-Horse Fertilizer-Grain Drill is of the same general construction as the plain five-disk drill, with the exception that it is equipped with combination fertilizer and grain box.

This drill distributes any standard fertilizer just as thoroughly as it plants seed. A lever in the grain box regulates amount of seed planted per acre, and a lever at the side of the box regulates the amount of fertilizer distributed.

Positive force-feed. The Van Brunt improved star finger feed has a capacity of 60 to 1070 pounds of fertilizer per acre, with but one change of speed. Agitators can be furnished for either grain or fertilizer.

Lifting lever at the back makes this drill easy to operate. The average farm boy can handle the lever. By pushing down on this lever, the disks are raised, and the drill is carried on the three wheels. When more pressure is needed to force the disks into difficult soil, the caster wheels can be raised by raising the lever, and the full weight of the machine carried on the disks.

Front View of John Deere-Van Brunt Model X Five-Disk Combination Fertilizer-Grain Drill.

B Advertising literature for the same drill

C Wooden-box 16-row Model "F" grain and fertilizer drill, which always had 7" spacing

D The "F" was replaced with the all-steel box "FF" drill, seen here behind a styled Model "B" tractor in May 1941

E A newly arrived "FB-B" 15-row drill at the author's farm in 1960

F The same drill sowing winter wheat that fall

Fertilizer Distributors

The original Van Brunt lime and fertilizer sower with the agitators on the axle had two other models added to the line in the early thirties. The "H" fertilizer distributor was followed by the "A" with star force-feed distribution.

In the mid-fifties great accuracy in placement was subsequently achieved with the tractor-drawn "LF," available in 8′, 10′ and 12′ widths. It was the writer's enthusiasm for this machine— a 10′ model was imported for the farm—which resulted first in their importation into and subsequently their manufacture in the U.K. The mounted equivalent, the "MLF" for liquid fertilizer, was added in due course.

B An 8′ Van Brunt lime and fertilizer sower on optional steel wheels in a 1924 catalog

A An old Van Brunt broadcast seeder on the Deere & Company display floor in 1987

C A similar wooden-box Model "H" sower on rubber tires as early as 1928

Spreading Calcium Chloride in Dodge County, Wisconsin, with a John Deere-Van Brunt All-Year Road Maker.

John Deere-Van Brunt Lime and Calcium Chloride Distributor or All-Year Road Maker

Equipped for Spreading Calcium Chloride, Sand and Ashes on Roads

The John Deere-Van Brunt Lime and Calcium Chloride Distributor, when equipped for road work, as shown on this page, is called the All-Year Road Maker. It has proved to be the most satisfactory machine for distributing calcium chloride on unpaved roads to settle the dust and for spreading sand, ashes or cinders on ice-covered highways to prevent skidding. Its use is advocated by manufacturers of calcium chloride because of its simple operation and even distribution—a requirement of first importance in applying the chemical salt.

The special equipment consists of a center shift lever and auto truck wheels that take pneumatic tires. The box covers are removed so that the material to be distributed can be shoveled from the truck into the box while the outfit moves along.

Rear view of John Deere-Van Brunt Lime and Calcium Chloride Distributor or All-Year Road Maker with pneumatic tires and center Shift Lever; less box covers. Revolving agitators are regularly furnished. Equipped with Rotary Wing Feeds, described on page 6.

D

D Another Model "H" sower for distributing lime and calcium chloride, but in this case adapted for road work and known as the All-Year Road Maker

E While the Model "H" had the agitators on the axle, the Model "A" shown had star-type force feeds to distribute fertilizer in quantities from 48 to 4,950 pounds per acre

E

Dusters and Ammonia Applicators

Two other types of machines which should be mentioned here were the various dusters offered, including the "DDT" applicator, and the 930 anhydrous ammonia applicator. The latter machine led the way to a line of tank and drum sprayers in 1959.

A The mounted version of the "LF," the "MLF," shown on a 630 tractor in 1958, has the optional grain box and grass seed attachment as well. It was offered in 6' and 9'3" widths.

B The very accurate "LF" distributor is seen here behind a 720 tractor and "CC" cultivator on rubber, and is followed by a spike-tooth harrow. Its rear drawbar allowed this "train." The "LF," which could apply fertilizer from 20 to 10,000 pounds per acre, is also equipped with the optional grass seed attachment.

C The 930 NH3 applicator side-dressing five rows of corn spaced 40" apart. The tractor is a 60 with wide front axle.

D A 620 tractor with "DDT" applicator mounted, in a July 1957 corn crop

Haying Review

The John Deere way of making hay had become an accepted way of life for grassland farmers. The original Dain idea, windrowing the mown grass while green into fluffy rows to allow the air and sun to cure stems and leaves more equally and thoroughly, gave the best feed.

Mowers

First, of course, the grass had to be cut, and Dain produced both a Plain Lift and a Vertical Lift mower for the purpose. The former was available with 5′, 6′ or 7′ cut, while the latter was only built as a 5′ cut machine. These early mowers had external drive gears. Deere acquired the company in 1910 and a new mower was introduced, the No. 1. The No. 2 Big Frame mower followed at the end of the decade, and in due course these were replaced by the No. 3 and Big 4 respectively. This latter stayed in the line into the fifties, by which time it was usually supplied on rubber tires.

[B] The early Dain mower from the Deere historical collection

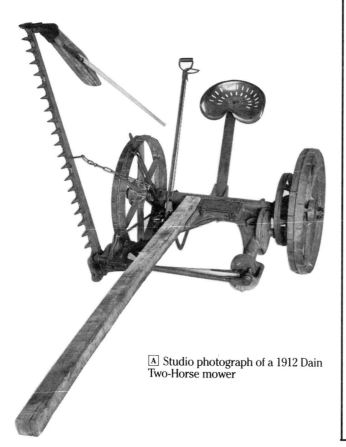

[A] Studio photograph of a 1912 Dain Two-Horse mower

[C] A restored No. 1 mower on the Deere & Company display floor in 1987

JOHN DEERE Enclosed-Gear MOWER
...the Mower That Oils Itself

Outstanding Features

Enclosed gears running in oil. Insure lighter draft, smoother and quieter operation and longer life.

Direct transmission of power through two pairs of perfectly balanced gears lightens draft and reduces wear. Thrust of each pair counteracts thrust of the other pair.

The heavy, one-piece axle has no holes to weaken it. Main drive gear and pawl plates splined to axle.

Two steel-encased leather oil seals on axle line and one back of flywheel prevent leakage of oil.

Durable, smooth-running clutch insures instant starting of sickle.

High, easy foot lift and extremely high hand lift furnish great clearance under all operating conditions.

Simple, effective adjustments for realigning cutter bar and for recentering knife, should this be necessary after years of use, add years to life of mower.

Cutter bar and knife carefully fitted. Hardened steel wearing plates and knife holders keep knife cutting true longer.

Hardened knife head guides and knife head wearing plates at both front and rear of knife head are readily replaced. Knife head extra long and reinforced.

Connection of cutter bar to extra-wide yoke is unusually strong. Two lugs on shoe at both front and rear, with hardened pins and bearings machined to size, keep cutter bar in line.

Latchless tilting lever is stronger and easier to operate.

Strongest possible construction of drag and brace bar—pitman protected.

Cutting parts made of best materials; improved construction increases cutting efficiency and reduces repair expense.

Flexible bar conforms to uneven ground. Powerful lifting spring causes bar to float, which increases traction and lightens the draft.

Pole protected under frame—easily removed without disturbing other parts.

Easy to repair—pawls, gears and shafts quickly removed with ordinary tools.

Simple, strong, durable and light draft. Easy to operate and adjust.

JOHN DEERE No. 3
Made in 3½-, 4½- and 5-Foot Sizes

D November 1935 advertisement for the No. 3 mower which replaced the No. 1

E Nos. 2 and 4 mowers in a sale at Sully, Iowa, on July 31, 1987

F The No. 4 with 2-horse hitch, stayed in the line until well after World War II. Shown here in a 1935 advertisement.

Power-Driven Mowers Introduced

The company's first power-driven mower was designed for the "GP" and simply referred to as the Tractor Mower. By 1934 the Trailer-Type Power mower was in fact the forerunner of one of perhaps the most famous worldwide, the semi-integral PTO No. 5 mower with 4½' to 7' cut. For the 1958 season, the new No. 8 replaced the long-lived No. 5 as the company's semi-integral model, and the No. 9 took over the fully mounted mower role. A new 10 pitmanless side-mounted mower was offered for the last of the Waterloo 2-cylinder tractors, and in their last year both the No. 8 and No. 9 were optionally available as the Nos. 8-W and 9-W for wide-tread tractors.

The smaller Dubuque tractors were originally matched by the 51 rear-mounted mower, but early in the fifties the M-20 center-mounted was introduced, stayed in the line for the 40 and 420 series tractors as the 20, and in 1957 was updated to the 20-A.

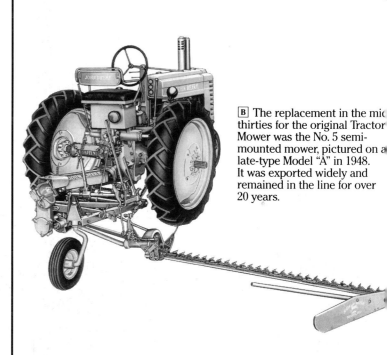

B The replacement in the mid thirties for the original Tractor Mower was the No. 5 semi-mounted mower, pictured on a late-type Model "A" in 1948. It was exported widely and remained in the line for over 20 years.

A Deere's first PTO-driven mower was the Tractor Mower, designed for the "GP" and shown in this 1929 photograph

C The No. 5 mower was eventually replaced by the No. 8, seen here in July 1957 with one of the new hay conditioners behind a 520 tractor

D When the No. 8 replaced the 5 a fully-mounted No. 9 was added to the line and is seen here on a 420U utility tractor

E The final design produced during the period covered was the mid-mounted 10, shown in August 1959 on a 530 tractor.

F The mower for the smaller tractors was the 20, here mounted on a Model "M"

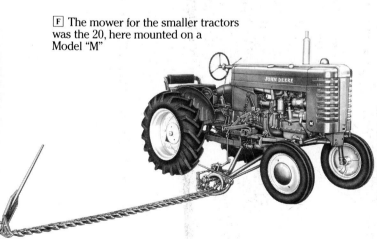

G The replacement for the 20 was the 20A, seen center-mounted on a 430 Row-Crop Utility tractor

Rakes

After the hay crop was cut, the next critical step was early windrowing with a Dain-type side-delivery rake. The slanting design of the frame and its curved teeth were the secret of this rake's success. Originally a single rear caster wheel was standard, but two were later adopted. For combination raking and tedding, another model used straight tines which could be angled for different crop conditions.

By the mid-fifties the 350 fully mounted or semi-integral PTO-driven models were joined in the line by the 851 ground-driven semi-integral rake and the new 594 and 594LW (low-wheel) versions of the original sloping-frame models.

Other machines introduced to speed up the haymaking process were hay conditioners and hay fluffers, while a sulky rake used originally as the primary rake before side-delivery models were introduced, but latterly for tidying up after all other operations were complete had been the first harvesting tool introduced by the company in the previous century.

B The original Dain side-delivery rake with single rear wheel

C The inclined-frame John Deere Dain system rake, shown in a 1922 advertisement with two rear wheels

A Dain side-delivery rake at work in a hayfield behind two horses in June 1936

D An updated version of the Side-delivery rake, the 594, pulled by a 430 Row-Crop Utility tractor. These rakes featured curved tines and a floating cylinder to ensure conforming to ground conditions.

E The final model of this line of rakes, the 594LW low-wheel version on rubber tires, with an "MT" tractor in the summer of 1950

F A new concept in rake design, the 851 featured either semi-integral or drawn types as here, with a new right-angle reel, ground drive, curved teeth, and wheel and reel in line, thus reducing the forward movement of the hay by 50%. The tractor is a 320 Standard model.

G The 3-point or semi-integral 350 side rake had the same reel as the 851 but was PTO-driven, in this case from a 530 in 1958

A Capable of raking 9′ swaths, the 896 integral ground-driven rake could work at speeds up to 8 mph. The tractor shown, a 530 with wide front axle, was difficult to find pictured.

B A 435 diesel tractor with a combination No. 9 integral mower and hay conditioner in a good alfalfa crop in 1959.

C The introduction of hay conditioners helped greatly in the curing of legume crops, with their juicy stems. A 430T-W wide- front-axle tractor is featured in this June 1959 photo.

D The end of the 1959 season saw the introduction of another hay tool, the No. 2 swath fluffer. Tractor is a 430RCU.

E A horse-drawn sulky rake, to tidy up after the hayfield was cleared, was the first harvesting tool introduced by the company in the 1880s. At that time it was the chief raking machine.

Sweeps and Stackers

Dain's first idea was to sweep the cured hay to a stacker, where it was put into a rick or stack. A number of different types and sizes of sweeps are shown with modified fittings as tractors gained greater flexibility. Stackers too were of various types, although the overshot version lasted the longest.

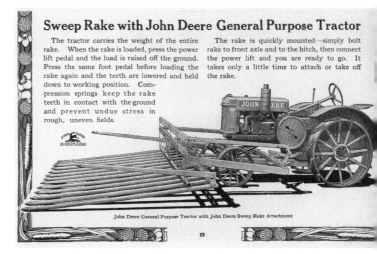

Sweep Rake with John Deere General Purpose Tractor

The tractor carries the weight of the entire rake. When the rake is loaded, press the power lift pedal and the load is raised off the ground. Press the same foot pedal before loading the rake again and the teeth are lowered and held down to working position. Compression springs keep the rake teeth in contact with the ground and prevent undue stress in rough, uneven fields.

The rake is quickly mounted—simply bolt rake to front axle and to the hitch, then connect the power lift and you are ready to go. It takes only a little time to attach or take off the rake.

John Deere General Purpose Tractor with John Deere Sweep Rake Attachment

13

B A 1930 illustration of a sweep rake mounted on a "GP" tractor and raised by the tractor's mechanical lift

JOHN DEERE PLOW COMPANY

Dain Senior Hay Stacker

A From a 1912 catalog, the Dain Senior hay stacker, capable of elevating hay up to 30′

C In August 1929, a "GP" with No. 1 mower attached operates a No. 2 stacker, capable of building 30′ stacks

D A charming photo of the John Deere Dain Automatic stacker at work in 1925, operated by two horses with a young lady in charge

Loaders

The alternative to the sweep-stacker operation was to load the hay on 4-wheel wagons, and for this purpose three different types of loaders evolved. The first of these—the Dain rake-bar loader, either 6′ or 8′ wide—had a double raking action on the ground, and was referred to subsequently as the John Deere-Dain Direct-Drive loader. It adjusted automatically to an uneven surface, as the rake bars were hammock-swung.

If one normally intended to load from the windrow the new Deere Single-Cylinder loader was the answer. Its cylinder was on the axle, and the slatted carrier handled the hay very gently, making it ideal for alfalfa, clover or beans.

Similar, but with a second cylinder at the rear to collect any crop missed by the first, was the Double-Cylinder model. Other variations were the green-crop loaders of two different types for uncured grass silage. One of these, the 204, was a heavy-duty green-crop loader for crops such as beans and peas for canners. With a slat-and-canvas elevator it could load hay or straw equally well. The 306 was the more conventional raker-bar type, also with a second raking cylinder for clean pickup.

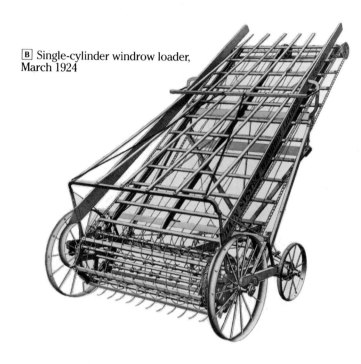

B Single-cylinder windrow loader, March 1924

A Dain rake-bar hay loader in July 1920

C A May 1939 picture of the more usual rake-bar type green-crop loader with a Model "H" and 20 trailer.

D A 1927 advertisement for a double-cylinder hay loader

E The strongest loader in the Deere line, the No. 204 heavy-duty green-crop loader. Designed originally to handle green lima beans and peas for canners, it had double drive from the 5″-wide rear wheels. Extension rims, lugs and wheel weights, or optional rubber-tired wheels could be used.

F One of the late-type loaders equipped with rubber tires, pictured loading a heavy crop of alfalfa in 1948

265

Balers

Another early alternative to stacking was to sweep the hay to a stationary press, which could be belt driven from the tractor, or fitted with its own motor. John Deere hay presses had a unique eccentric gear drive which gave the plunger extra power and also meant that the donkey head had a very rapid withdrawal from the bale chamber after each charge of hay or straw. As early as the First World War three sizes of these stationary presses were in the company's catalog.

B A 1929 factory photo of a Dain Pull-Power press

A A nicely restored Power Hay Press on the Deere & Company display floor in 1987

C A 2-horse Pull-Power press with its feed chute folded, in a sale at Le Grand, California, on September 26, 1986

D

E

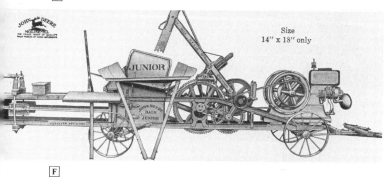

F

D Motor press with mounted stationary engine at work in 1922. This model was available in 14″ × 18″ and 16″ × 18″ sizes.

E This August 1918 factory photograph shows a Junior hay press with a recently acquired Waterloo Boy engine in place of the earlier Root and Vandervoort model. This picture clearly illustrates the John Deere eccentric drive gears used on all their presses before World War II.

F Right side view of Junior press with a different-style Waterloo Boy engine mounted, and 2-horse shafts. The Junior was only available in 14″ × 18″ size.

Windrow Pickup Press

It was the year 1936 which saw the biggest step forward, with the introduction of the windrow pickup press. This was fitted originally with a Lycoming engine, later replaced with a Hercules 4-cylinder motor. This machine eliminated at a stroke the rather wasteful field operations, picking up the hay direct from the windrow much as the earlier loaders had done.

Used extensively in both North America and in other parts of the world, these early pickup balers became increasingly popular, even though they were still labor intensive and used largely by custom operators. The demand for a smaller machine for more general farm use saw the introduction in 1942 of a PTO-driven machine of reduced size.

Automatic Wire-Tying Pickup Balers

With the high labor requirement of both the above machines, it was imperative that automatic pickup baling should be perfected. Deere's answer in 1946 was the introduction of the 116-W automatic wire-tying baler.

After initial problems in the first season with movement between the twisters in different hay conditions—causing all 3500 units to be returned to the factory for modification during the first winter—the 116-W was a very successful baler, and was exported worldwide. The initial model made 16″ × 18″ bales, but a smaller version, the 114-W, was marketed from 1953 on.
It made 14″ × 18″ bales. The original problem had been overcome by mounting both twisters in a cast cradle.

A The smaller power take-off pickup press, designed for the average farmer, featured a ground-driven pickup and the usual Deere eccentric gears

B A 116-W hay baler pulled by a late Model A tractor is shown in a July 1952 photo

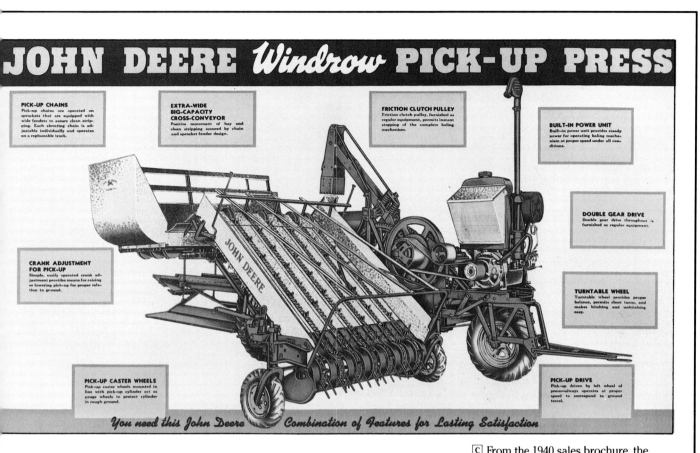

JOHN DEERE *Windrow* PICK-UP PRESS

PICK-UP CHAINS
Pick-up chains are operated on sprockets that are equipped with wide fenders to assure clean stripping. Each elevating chain is adjustable individually and operates on a replaceable track.

EXTRA-WIDE BIG-CAPACITY CROSS-CONVEYOR
Positive movement of hay and clean stripping assured by chain and sprocket fender design.

FRICTION CLUTCH PULLEY
Friction clutch pulley, furnished as regular equipment, permits instant stopping of the complete baling mechanism.

BUILT-IN POWER UNIT
Built-in power unit provides steady power for operating baling mechanism at proper speed under all conditions.

CRANK ADJUSTMENT FOR PICK-UP
Simple, easily operated crank adjustment provides means for raising or lowering pick-up for proper relation to ground.

DOUBLE GEAR DRIVE
Double gear drive throughout is furnished as regular equipment.

TURNTABLE WHEEL
Turntable wheel provides proper balance, permits short turns, and makes hitching and unhitching easy.

PICK-UP CASTER WHEELS
Pick-up caster wheels mounted in line with pick-up cylinder act as gauge wheels to protect cylinder in rough ground.

PICK-UP DRIVE
Pick-up driven by left wheel of press—always operates at proper speed to correspond to ground travel.

You need this John Deere Combination of Features for Lasting Satisfaction

C From the 1940 sales brochure, the windrow pickup press is shown in its final form with a Hercules power unit and on rubber tires

D Studio photograph of a 114-W baler

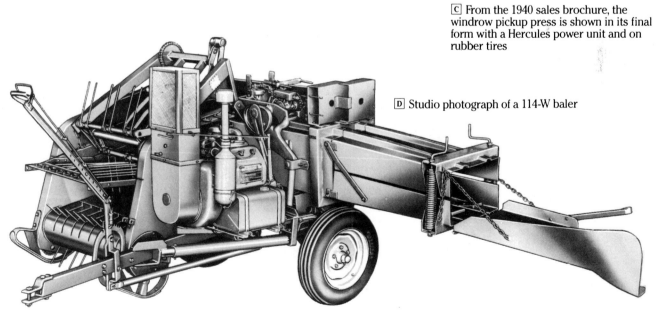

269

Automatic Twine-Tying Balers Added

It was 1954 before an automatic twine-tying pickup baler, the 14-T, was added to the line. This machine stayed in production into the sixties, when it was superseded by the 24-T, but it had set the design pattern for all Deere's subsequent square balers with its auger-fed bale chamber.

In the meantime, demand for a larger baler saw the announcement of the 214-T, first in the U.S. in 1957 from the Ottumwa Works, and in the following year in Europe from the recently acquired Lanz factory in Zweibrucken, West Germany. These were the first Deere machines available from the European factories following their purchase. An alternative wire-tying machine, the 214-WS, took the place of the popular 116-W at the same time. Both of these balers had a twisting mechanism which did not deposit small cutoff pieces of wire on the ground, as did some of the competitors, to the danger of livestock.

For the very large farmer or custom operator, the final development in the 1950s was the introduction of the 323-W 3-wire baler fitted with a 4-cylinder Deere auxiliary engine made in Dubuque and capable of producing bales 16″ × 23″ and up to 50″ long. These balers were particularly useful in California and other southwestern states, where alfalfa hay grown in the south had to be transported to the northern cattle farmers.

A A 420U utility tractor and 14-T baler make light work of a poor hay crop

B This 14-T baler is powered by a 50 tractor and is fitted with the optional bale ejector

D Studio photograph of a 214-WS baler

C A 530 tractor in charge of a 214-T pickup baler in August 1958

E Deere's largest baler in the fifties was this 323-W 3-wire pickup baler, seen here with a 4-cylinder Dubuque engine and pulled by a single-front-wheel 730 diesel tractor

Forage Harvesters

In 1936 the company introduced its first ensilage harvester for use in corn. These machines, and the introduction of machinery for loading and emptying tower silos, led inevitably to the idea of chopping grass for silage. In 1942 the 60 series of flywheel-type forage choppers was announced, the 62 for pickup work, the 64 with a corn head, and the 66 with cutter bar for cutting the crop direct.

B The company's first ensilage harvester designed for corn, being used with a raker-bar green-crop loader to chop and load alfalfa in July 1937

A By the mid-forties the 60 series of field choppers were developed and the 1945 Better Farming full-line catalog featured a 62 on its cover page. This 1948 scene shows one operated by a late-type "A" tractor.

C The field chopper fitted with corn head was the Model 64

John Deere

THE John Deere No. 66 shown here is equipped for field harvesting of grass silage. Scene at lower right shows the No. 66 mowing and chopping alfalfa for silage.

Page 32

D The 66 was equipped with cutter bar to mow and chop alfalfa and similar crops direct

E No. 8 forage harvester chopping corn in 1952 pulled by a 60 tractor

Forage Harvesters Updated

The 60 Series forage harvesters proved to be very successful and remained in the line for a number of years before their replacement with the No. 8. It in time gave way to two machines, the lighter No. 6 and the heavy-duty 12. Both were offered with pickup or corn header options, and on the No. 6 a 4', 5' or 6' cutter bar while a 6' or 7'-cut unit was available for the 12.

Another method of cutting and loading grass and other crops was with the double-chop machines developed by Lundell and built by Deere to Lundell's patents. The first of these was the 10 with rotating flails to cut the crop, which was then fed into a flywheel blower equipped with knives to further cut the material. These machines achieved quick popularity with farmers, and the 10 gave way to an improved version with greater capacity, the 15, in 1958.

B One of the forage harvesters replacing the No. 8 was the No. 6, seen here behind a 630 tractor and with a 110 Chuck Wagon in tow in June 1958

A A 720 diesel tractor easily handles a No. 8 forage harvester and "N" spreader with forage box attachment in a heavy mid-west crop

C A similar outfit but with the direct cutter-bar head on the forage harvester

D For heavier duty the other machine in the line was the 12, seen here with a 730 tractor and a truck-mounted 110 Chuck Wagon alongside

E In June 1959 a 630 tractor powers this 7′ direct-cut 12 forage harvester

F The engine-mounted option on the 12 forage harvester allows a 630 tractor to cope with this heavy corn crop and tow a 110 Chuck Wagon, in September 1959

G This 620 and 15 double-chop tows an "N" spreader with forage box attachment in July 1957

Forage Blowers

To deal with the resulting material from flywheel forage harvesters, and double-chop rotary choppers, forage wagons were developed with a conveyor-type bed and front cross-conveyor to move the transported material into blowers that could fill tower silos. Early model blowers were introduced in 1936-37, and refined versions—the No. 2, 50 and 55—followed up to the late fifties.

In addition to the popular but specialized 110 Chuck Wagon, Deere offered an attachment for its largest capacity Model "N" spreader which converted it into a forage wagon, thereby making it a dual-purpose machine.

B The 50 blower featured auger feed and retracting wheels to keep the machine stationary when positioned

A One of the first silage blowers, pictured at the factory in 1944

C The replacement for the 50 was the 55 conveyor-fed blower

D This 110 Chuck Wagon is in use behind a 12 forage harvester and 630 tractor

E An excellent illustration of the operation at the tower silo. A 620 and 110 Chuck Wagon unload into the hopper of a 55 blower, which is driven from another 20 series tractor.

F Another use for a forage box, in this case the "N" spreader attachment, unloading from the side into feed bunks

Harvesting Review

Grain Binders

The company's real involvement with the grain harvest began in earnest in 1910 with its decision to build a binder—or harvester as they were often referred to in those days—to compete with the already popular McCormick-Deering and Massey-Harris machines, to name but two. The opening in 1912 of the new Harvester Works in East Moline, next to the existing Marseilles Spreader Works, enabled this decision to be fully augmented.

The Light-Running John Deere binder achieved its own degree of popularity and remained in the line for many years. In due course, tractor power took over from horses or mules, but the binder remained ground-wheel driven until the introduction of power take-off on tractors made it possible to introduce PTO-driven models. These were offered with a 10′ cut initially, but the introduction of smaller tractors made an 8′-cut option popular.

Another development with tractor power was the introduction of remote controls enabling the tractor operator to sit on the binder, making a one-man outfit possible. An alternative was the multiple hookup with two or more binders hauled by one tractor. Up to six were arranged with some of the very large tractors used on the Great Plains and the Canadian Prairies early in this century.

Shock Sweeps

After the sheaves were cut and tied with a binder, they were placed by hand in shocks to dry out. This operation was assisted by the use of a sheaf carrier available as an extra on the binder. With first mechanical and later hydraulic lifts introduced, a grain shock sweep was added to the line.

A A 1915 Harvester Works advertisement for the Light Draft horse-drawn ground-driven grain binder

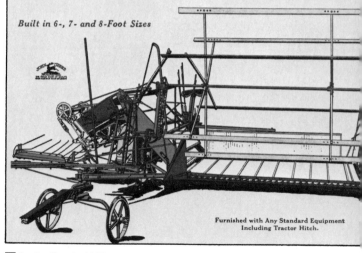

B In the line in 1930, the improved Light Running grain binder for horses or tractors was offered in 6′, 7′ and 8′ sizes

C This 10'-cut ground-driven tractor binder requires four horses to pull it

D Illustrating a very unusual occurrence, the "D" is pulling two 10' PTO-driven binders, the second obtaining its drive through the first

E A power-driven 10'-cut binder behind an "A" tractor in 1937

F This 1937 advertisement offers the tractor binder in 8' as well as 10' size

A After the binder sheaves had matured in shocks a shock sweep was a great help in collecting the crop and delivering it to an elevator for stacking

B This shock sweep mounted on an "A" tractor is delivering its load to a thresher

C This PTO-driven corn binder has the standard power carrier to place the bundles clear of the Model "B" tractor's next pass

D The grain binder's equivalent for corn, advertised in 1915

E A power corn binder with optional elevator attachment and trailer side hitch

Corn and Rice Binders

During the First World War, a corn binder had been introduced, first horse drawn and later tractor drawn as with its grain counterpart. Later with PTO drives available, special binders for use in the rice-growing areas also became possible.

A In September 1937, a "D" tractor with extension rims on both front and rear wheels copes well with this PTO-driven rice binder

B Three power-driven rice binders in the 1937 harvest with an "A" and two "D"s, all equipped with rear extension rims and electric headlights

C A styled "D" and PTO-driven rice binder in a good 1939 rice crop

Deere Experiments with Combine Harvesters

In 1925-26, the company decided it must consider the gradually increasing popularity of the combine harvester. The possibility of purchasing a company already making combines was considered, but after trials with two other makes, compared with an experimental machine built in the Harvester Works, Deere found that its own machine performed equally well, and therefore decided to build its own.

Production Commences

In 1927, there were 50 of the No. 2 combines at work, followed the next year by the smaller No. 1. The former was available with 12′ or 16′ platform and a 40″ wide separator, while the latter offered the options of 8′, 10′ or 12′ cutter bar. Both models had a 24″ × 22″ spike-tooth cylinder and built-in recleaner to give a good finished sample of grain.

B One of the prototype combines of the mid-1920s has the Marseilles (Spreader) Works as a background. The offices at the end of the building now house the Deere Archives, source of many of these beautiful pictures.

C John Deere No. 2 combine, the first to appear in 1927 with either 12′- or 16′-cut and a 35-hp motor. The cylinder was 24″ wide and 22″ in diameter, and the separator was 40″ wide.

A An early attempt at a whole-crop harvester with one of the first 50 Model "D" tractors

D A factory photograph of the No. 2 in 1928

E The No. 1 model announced in 1928 was available as 8'-, 10'- or 12'-cut. Otherwise it was a smaller version of the No. 2

Study This Internal View of the John Deere Combine

F An internal view of the No. 2 showing the spike-tooth cylinder, four straw walkers, built-in recleaner and optional straw spreader

Developments in 1929

In 1929, the No. 3 replaced the No. 2; it had been made lighter as the market and conditions demanded, but had a 30″ × 22″ cylinder.

Experiments with a hillside combine, the No. 4, were made in 1929, but were halted without the machine going into full production when Caterpillar offered Deere their Western Harvester (Holt) combine line that year. The price asked at the time proved too high, but the offer caused delay in introducing a hillside model until 1936, when an alternative suggestion proved acceptable.

A A three-quarter rear view from the sales brochure of the No. 3 combine, which replaced the No. 2 in 1929 and was of lighter construction

B The No. 3, fitted with a rotary pickup platform, and Model "D" tractor in 1929

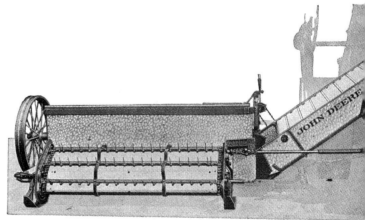

C The raddle-type pickup shown compares with the previous photo's rotary type on a No. 3 combine

D Front view of No. 3 combine in a Kansas wheat field in the thirties

E Another view of the same outfit hauled by a Model "D" on steel wheels with extension rims

Continuous Model Update

In the meantime the No. 5 replaced the No. 1 in 1929 and in turn was updated to the No. 5-A in 1934. These machines had a 3-walker 30"-wide separator and a 50-bushel grain tank, compared with the 40" width and 65-bushel capacity of the No. 3. The new large 17 that replaced the No. 3 in 1932 had similar dimensions except that the separator width was reduced to 36".

The Nos. 2, 3, and 17 models had a 35-hp Hercules motor, while the Nos. 1, 5, 5-A, 6 and 7 had a 24-hp Lycoming.

B The production No. 5 photographed in April 1930 in the Harvester Works with rotary-type pickup attached and the motor in the normal front position. On top of the tank is the optional weed-seed cleaner.

A A preproduction No. 5 in October 1929 with the motor mounted on top of the separator and under the tank

C Left-side view of No. 5 with sacker and grader outside the Harvester Works

D

E

F

G

D The No. 5 shown in its transport position

E A No. 5-A 12'-cut combine harvesting behind a "GP" tractor

F Model 17 tanker 16' combine equipped with weed-seed cleaner harvesting soybeans behind a Model "D" tractor in October 1935

G In May 1940, a rubber-tired No. 17 tanker combine with pickup attachment is harvesting a windrowed crop

A Small Model Required

The need for a small combine had become obvious in the early '30s, but it was 1936 before the No. 6 was introduced. It had its threshing mechanism crossways, as did the increasingly popular Allis-Chalmers 60, but unlike it, retained the spike-tooth drum of the larger models. Despite a cylinder and straw walkers the same size as the No. 7 8-foot-cut model which preceded it, the No. 6 was not a success.

The introduction of the 8' cut No. 7 in 1932, ahead of the 6' cut No. 6, gave the company their first one-man combine. This model, which remained in the line until 1941, had conventional layout, with 24″ × 22″ cylinder as on the No. 5-A but with 24¾″ separator width. It was replaced for the 1941 season by a newly styled No. 7A, but wartime restrictions caused the dropping of this model along with the No. 5A, 33 and 35.

A Cutaway picture of the No. 6 combine from its sales brochure showing the spike-tooth cylinder and crossways straw walkers

B October 1936 sees a No. 6 combine harvesting soybeans behind a Model "A" tractor with 4-stud front pedestal and French & Hecht wheel equipment

C A recent find—rear view of a No. 6 combine with engine drive, found in a field near Buxton, North Dakota, and now the property of Louis Toavs of Wolf Point, Montana, possibly one of the largest private collectors of old tractors and combines in the world, all John Deere

D A December 10, 1940, Harvester Works photo of the then new No. 7-A combine

E Note the detail changes in this July 29, 1941, picture of the No. 7-A—the main grain elevator has been shortened, the rear hood is deeper and the header divider and radiator air intake are different

F A front view of the same machine on the same day

G The front cover of the No. 7-A sales brochure shows the new combine at work

JOHN DEERE
No. 7-a
COMBINE

Cuts an Eight-Foot Swath—

Power Take-Off or Engine Drive Is Optional Equipment

When you buy John Deere Implements you are sure of prompt repair service during their long life.

Holt-Caterpillar 36 Acquired

The year 1936 was more famous as the year when the Caterpillar Model 36 joined the line, with all the history acquired by this decision. Holt had become the largest producer of combines in California, having gradually absorbed almost all the competitive makes, most of them in the 1890s but finally their largest competitor, Best, in 1912.

Over the years the Holt company achieved many firsts. The first sidehill combine appeared in 1891, and a gas-engine-driven model in 1904. This led to development of the first full-production self-propelled machines in 1911.

Enormous combines with cutter bars up to 44′ were marketed—a 50′, cut combine, No. 574, was tried but proved too unwieldy.

A Benjamin Holt

B Horse-drawn medium-size Holt combine, ground-wheel driven, with 19 horses, many of which are interested in the cameraman. Sometimes two but as here three lead horses were controlled by the operator, followed by up to five gangs of six horses each.

C

F

D

E

C Haines-Houser No. 583, the second oldest known preserved combine, pictured in the Stockton Museum on September 24, 1986. Holt purchased Haines-Houser in 1902 but kept their combines in production until 1914.

D Holt's first small sidehill combine at work on Frank Hardin's ranch at Davenport, Washington, in 1898. Man at lower left is Pliny E. Holt, nephew of Benjamin Holt.

E Holt steam engine No. 25 in charge of an early Holt combine fitted with grain lifters and a heavy-duty reel

F The world's first self-propelled combine, built in 1886 and used on the farm of G.S. Berry. It had 22′ cut and harvested 50 acres of grain a day. It was propelled by the first straw-burning steam engine and was later fitted with a 40′ header, allowing it to harvest, for the first time, over 100 acres a day.

A An original-condition Holt gas-engine-driven combine in what was probably the world's largest farm machinery sale at Le Grand, California, in September 1986

B A similar restored combine makes an interesting comparison with the previous picture

C One of Holt's largest combines, its Standard Steam combine, with dual wheels and a 12′ extension to the basic 22′ cutterbar

D A 1915 photo of a Holt 16′ self-propelled combine owned by John C. Evers of Almira, Washington

E Holt's Baby self-propelled combine, pictured about 1916 at Holt's Stockton Harvester Works

A Full Line

Some eight different models were listed in Holt's 1912 catalog, ranging in size from the Baby Special to the Large Standard. In 1919 came the introduction of the first all-steel machines, and this Model 30 is represented here by the well-preserved example displayed in the Lodi museum near Stockton, California.

In 1926, the 30 was superseded by the 36. This famous model, marketed first by Western Harvester, the Caterpillar subsidiary, as the Holt 36, then from 1930 as the Caterpillar 36 and finally by John Deere in 1936 as their Model 36, stayed in production until 1953, longer than any other combine.

B The only known picture of Holt's 50′-cut combine, No. 574, which was too wide to be manageable; built in 1893

A Holt Baby gasoline-powered sidehill combine with Waukesha motor, built in 1921 at Spokane, Washington, by a Holt subsidiary, Northwest Harvester Co.

C One of the largest combines built by Holt

Holt Caterpiller & Combined-Harvester,
run by David Jasman and family
near Wheeler, Wash., Aug. 29, 1913.

[D] Holt Caterpillar and combine harvester, run by David Jasman near Wheeler, Washington, on August 29, 1913

[E] Model 30 20'-cut all-steel combine at Lodi Museum near Stockton, California. Holt introduced the all-steel combine in 1919.

[F] Another Model 30 with sacking platform and weed-seed cleaner on the final clean-grain elevator, photographed at the Stockton Works

Hillside Combine Models Multiply

Eventually painted green and yellow and available on rubber tires, the Model 36 is shown well in the accompanying photograph. As mentioned earlier in the book the two smaller Caterpillar models, the 34 and 38, which competed to a degree with existing John Deere models, were not part of the Caterpillar-John Deere deal.

As a result, Deere first produced in 1937 a slightly smaller version of the 36, the 35. In 1940, another hillside model called the 33 was added, based largely on the new No. 9 model and with similar Dreyfuss styling.

B Left-side view of a Holt 30 combine, still displaying the decorative lining from the wooden era

A An extreme-hillside 36 bulk tanker combine on trial in 1928, from the Caterpillar subsidiary, Western Harvester Co.

C Three-quarter rear view of the same machine showing the heavy beams used to counterbalance the header, and the right-side sack chute

D Holt's Model 36 Level-Land tanker combine with 20′ cutter bar in its final pre-Deere form

E Near Dixon, Calif., in 1929 a Model 36 with pickup header collects windrowed grain flattened by north winds. At the rear is a buncher for straw saving.

F Although a model not built by John Deere after the Holt-Caterpillar acquisition, this Holt 34 with pickup head, pulled by a Model "D" tractor on steel wheels with double extension rims in 1929-30, has its own charm.

A Three 36 hillside combines, Caterpillar hauled, work on the slopes in Western wheat

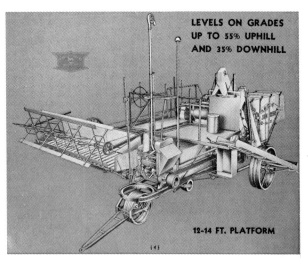

LEVELS ON GRADES
UP TO 55% UPHILL
AND 35% DOWNHILL

12-14 FT. PLATFORM

B The Model 35 was a smaller version of the 36 with 12' or 14' platform and is seen here in a picture from the May 1941 sales brochure. It was equipped with a 6-cylinder $3\frac{3}{8}'' \times 4\frac{1}{8}''$ Hercules engine running at 1400 rpm and had a $24'' \times 22''$ spike-tooth cylinder and 30'' wide separator.

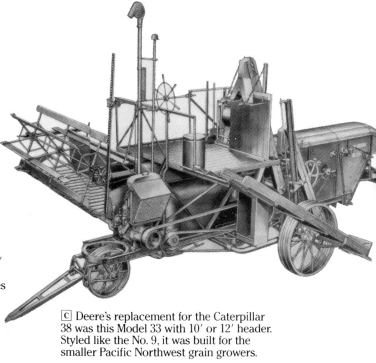

C Deere's replacement for the Caterpillar 38 was this Model 33 with 10' or 12' header. Styled like the No. 9, it was built for the smaller Pacific Northwest grain growers.

CASH *and* GRAIN SAVER *for* PACIFIC-NORTHWEST GRAIN GROWERS

⧈D⧈ This 33 is direct-hitched to a Caterpillar tractor to give greater control on hillsides. Both tractor driver and header operator are given protection from the sun; the combine is on rubber tires and is equipped with a straw spreader.

⧈E⧈ This 33 hillside combine on wartime steel wheels had a 4-cylinder 3¼″ × 4″ Hercules motor governed at 1800 rpm, and a 20″ × 22″ spike-tooth cylinder ahead of a 28″-wide separator

⧈F⧈ The 36-B shown here was the medium-hillside version of the 36 series, the 36-A being the extreme-hillside model. The 36s had a 6-cylinder 3¾″ × 4¼″ Hercules engine running at 1400 rpm, and the threshing cylinder was 30″ × 22″ with a 36″-wide separator.

Small Combines

Because the No. 6 was not a popular model, experiments with a straight-through series resulted in the announcement in 1939 of three small combines, the 10, 11 and 12, with 3½', 5' and 6' cut, respectively. All had right-hand cutter bars. It was soon realized that with most binders having left-hand cut, the two types of machines were not compatible.

B Four No. 9 combines loaded on a railroad car awaiting dispatch from Moline, Illinois, in December 1939

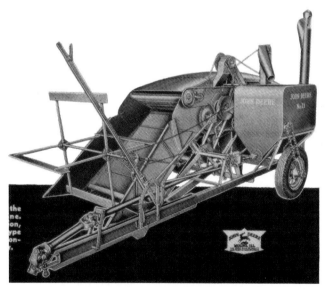

A From a May 1939 mailer, the 11 power-driven 5'-cut tanker combine, again with right-hand cutter bar

C The largest of the trio of small straight-through combines, the 12 uses grain lifters to pick up this windrowed crop. The Model "A" tractor has unusual twin rear wheels in this July 1939 picture.

D A unique photo of a preproduction No. 9 combine before the model was styled by Henry Dreyfuss. Shown behind a Model "A" tractor in November 1938.

E One of few pictures in existence of a 3½'-cut 10 combine with right-hand cut

Small Combines Updated

Accordingly after only one season these small models were changed to left-hand cut, and became the 10-A, 11-A and 12-A. These machines had 36″, 48″ and 60″ wide drums, respectively, all of 15″ diameter. Again the intervention of wartime restrictions caused the dropping of the 10-A and 11-A, although experiments were conducted in 1944 with a 10-A combine fitted with the 11-A 5′-cut platform and called the 10-AW. This model did not go into production.

The 12-A proved very popular and remained in the line until the updated 25, with optional 6′ or 7′ cut, took its place in 1952. This was replaced in turn with the auger-and-raddle-feed 30 in 1956, with only a 7′ platform available. Both these models retained the 12-A's 60″ × 15″ drum, and the latter two had their grain tanks enlarged, from 20 to 25 bushels.

B No. 3 experimental "M" tractor pulling a 10-AW combine (the 10-A with 11-A 5′ header) in July 1945 on the Henry Arp Farm, Moline, in a heavy crop of oats

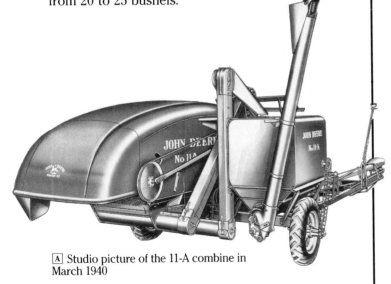

A Studio picture of the 11-A combine in March 1940

C Another view of the same outfit

D The small right-hand-cut combines were replaced after only one year by the left-cut 10-A, 11-A and 12-A models with similar cutting widths. The leading 10-A is pulled by an "H" tractor.

E In August 1951, a late-type Model "B" and PTO-driven 12-A combine at work in typical Midwest country

F Another late Model "B" pulls an engine-driven 12-A combine with draper pickup in a windrowed crop in October 1950

End of the Tractor-drawn Combines

To complete the story of tractor-drawn combines the 65 auger-header model, based on the 55 self-propelled, joined the line in 1949 to fill the gap left by the withdrawal of the No. 9. Both models had 12′-cut auger header and 30″ × 22″ cylinder, 30″-wide separator 130″ in length, and 45-bushel grain tank. The 65 stayed until larger trailed machines finally replaced it in 1966. It was particularly popular in grain-growing areas where windrowing was the practice, and the advantages of self-propulsion less evident.

B One year earlier a late-type "A" is in charge of a 65 combine in a crop of oats

A A later Model 65 combine behind a 620-W tractor. Note the differences with the above photo – auger feed to tank, narrow platform and folding unloading tank auger.

C The Model 12-A 6′-cut combine was finally replaced in 1952 by the 25, with either 6′ or 7′ cut and the option of PTO or 4-cylinder engine drive. This July 1952 picture shows the latter option behind an experimental "MTA" tractor, forerunner of the 40T.

D Front view of the same Model 25 in a nice crop of oats. The 25 retained the 12-A's cylinder and straw rack, but both sieves were lengthened by 2".

E Another Model 25 operated from the PTO of a 50 tractor cuts soybeans in October 1952

F The final version of the straight-through combine appeared in 1956 with the announcement of the Model 30. This combine retained the threshing mechanism of the 25, but used a retracting-finger auger and raddle to feed the cylinder. It was only available as a 7'-cut machine, but also kept the other options of the previous models.

Self-Propelled Combines: Experiments in 1944-45

During the war years experiments continued with self-propelled combines. The success of the Massey-Harris Harvest Brigade acted as a further spur, if one was needed. As we have already seen, a small self-propelled based on the successful 12-A was built and designated the "44" but did not go into production, though it could have proved very popular, but all effort was concentrated on the larger 12'-cut type, known all along as the Model 55.

Originally based like the 33 on the No. 9's threshing unit, with a sloping hood from the cylinder to the straw outlet, the 55 was tried with both canvas and auger headers, and with both sacking platform and bulk grain tank. The engine position in these first models was on the left side of the machine, in a similar position to the I.H. 123 and 125 combines of that period. Who copied whom I know not; even the left-side driving position was the same.

These models were also tried on crawler tracks for the rice fields. The next development—the 55X—was to have a horizontal hood line, the same header options, bagger or tanker, but with the tank and engine now on top of the separator, the latter beside the offset operator.

B The same machine on the same day, showing the side-mounted engine under the sacking platform and the grain grader

A This photograph (May 29, 1944) shows an early Model 55 combine with sacking platform, auger header, offset driver's position, and sloping separator housing like the No. 9 combine

C This May 24, 1944, Harvester Works photograph shows the experimental 55 combine with auger platform, tank attachment and side-mounted engine

D Another Model 55 at work with an extended radiator-air intake and an operator awning

E This picture shows the alternative canvas platform and the rear shelf for spare grain sacks

F A combination of canvas platform, tank attachment and crawler tracks make another interesting development

Design Finalized

Eventually the engine was moved behind the tank, adopting its final position, but the operator's platform remained offset. These combines, now called the 55C, were getting close to the final design production unit, where the driving position was centralized, offering a layout and threshing principle later adopted by many competitors.

In fact, the 55 was very advanced for 1946 when it finally went into production, and was more copied than any other combine before or since. The European Claeys was virtually the same machine widened in the separator from 30″ to 40″ for the heavy straw conditions prevailing over there. Their first Model MZ eventually became the Clayson M 103, and this model evolved into the New Holland M 133 and today's Ford-New Holland machines.

Similarly Germany's Claas self-propelled, offered first in 1953 as the SP 55, another close derivative, and many others of the market's most successful designs owe their origin to the 55. As W.E. (Bill) Murphy, or "Murph" as he was affectionately known worldwide, the Harvester Work's installer of combines from Moscow to Argentina to the U.K., told this writer on one occasion, the 55 was "right from the word go."

Rice and Hillside Models

Rice models on oversize tires (18-26 front and 7.50-16 instead of 5.50-16 rear), the 55-R, or on crawler tracks, the 55-RC, were offered as options to standard, and in 1954 a hillside version joined the line. Eventually a 55-B with pickup header and edible bean attachment rounded out the 55 line, and after the introduction of the 2-row corn head on the later and smaller 45 combine, this further option was adopted for the 55 also.

A By December 13, 1944, the machine is designated 55X and has acquired a level separator hood. The engine and grain tank have been moved to the top of the combine, the engine in front alongside the operator's platform, and all the drives have been simplified. The twin straw spreader is in the lower under-driven position.

B The 55C of April 26, 1945, is beginning to look like the production machine, although still fitted with a canvas header and side-position operator's platform. The cylinder drive is now by V-belt, as is the straw spreader, which is top driven now. This particular machine has flax rollers ahead of the cylinder.

C A three-quarter rear view of the same machine on the same day. This photograph shows the variable road speed and flat-belt drive to the threshing mechanism adopted for the production models.

D The hillside version of the series, the 55-H, tackles the sort of steep slope it was designed for. The date is September 1954.

E This November 1952 picture of the 55-RC Rice combine with crawler tracks in a heavy crop shows the machine with a large tank extension

F These two 55-R Rice combines with sacker attachments are both of the later type with angled steering wheel and a 64-hp John Deere engine in place of the 60-hp Hercules fitted previously. The crop condition is difficult, to say the least.

Production Commences

Illustrated is one of the first year 55 combines near Champaign, Illinois, in 1947. One of the two 1948 machines ordered for the U.K. during the writer's first visit to Moline in September and October 1947, is also shown at his works duly restored some years later.

Dimensions of the 55 were as for the 65 already given, with the addition of a Hercules QXD $3\frac{7}{16}'' \times 4\frac{1}{8}''$ 6-cylinder engine, double hydraulic lift rams for the platform, and initially a hand-controlled variable speed lever, later replaced by hydraulics. The first year's production had 9-24 tires on the drive wheels but these soon proved inadequate and were increased to 13-26 for the 1948 season. The 55's ability to unload its tank either moving or stationary in 1½ minutes added to its plus points with the farmers, and the centered design was a great help with the next development.

B One of the two Model 55s imported into the U.K. for the 1948 harvest, both of which came into the author's possession in 1953. This is No. 2,638. It was converted from a sacker to a tanker and restored at that time.

A The author enjoying himself on one of the first year's production 55s (1946) near Champaign, Illinois. One of the brothers who owned the machine is instructing the "new operator" in a soybean field.

C Left-side view of the same machine

D Cutaway view of one of the most popular of Deere's combines, the 55, from a 1951 sales brochure

E The inside story of Deere's other most popular combine, the 36, makes interesting comparison with the 55

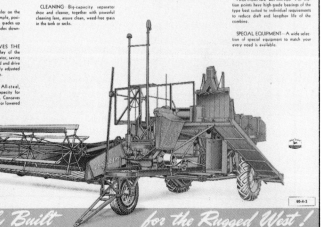

F Externally the 36 in its last form was painted green and yellow and was mounted on rubber tires all round. This page from a 1948 sales brochure illustrates the 36-B Hillside combine in this form with sacking attachment.

The Model 45 Announced

The 45 was a narrower version of the 55 but with a different frame construction, and it too became very popular. With 8′ or 10′ header, and later the 10 corn head, it was also available in rice and bean models called the 45-R and 45-B, respectively. It was not offered in hillside form, however, nor with crawler tracks.

Some of its specifications were 26″ × 22″ rasp-bar cylinder, 26″-wide separator the same length as the 55 and also with three straw walkers, 12-26 tires, and a Hercules JX4-C3 3¾″ × 4¼″ 4-cylinder engine. It was equipped with a 3-speed transmission instead of the 4-speed of the 55, and a 40-bushel tank.

A Baby Self-Propelled

Two other models should be mentioned before leaving the combine section. In response to requests from small grain and corn growers, the company announced the Model 40 at its large Marshalltown John Deere demonstration in 1959. This small self-propelled may well have filled the market niche originally intended to be covered by the "44" experimental machine some 15 years previously.

With a new 2-row corn head, the 205, or an 8′ or 10′ grain platform, the 40 was an attractive little machine which stayed in the line for seven years. A trailed PTO version, the 42, was to become available later. They featured a 24⅝″ × 22″ cylinder, 24⅝″-wide 110″-long separator, and 35-bushel grain tank. The 40 had a 42-hp John Deere 4-cylinder engine with 3-speed transmission.

A Deere's second production self-propelled combine was the Model 45, seen here in July 1957. Its layout was based on the 55, although its spot-welded frame construction was different. Available as an 8′- or 10′-cut, it was initially fitted with a 42-hp Hercules JX4-C3 engine, with 3¾″ × 4¼″ cylinders.

B In October 1958, another 45 with John Deere 145 engine cuts four rows of soybeans direct with its 12′ cutter bar

C The 45 at the Marshalltown John Deere demonstration in 1959 was equipped with a 2-row 10 corn attachment

D Announced at the Marshalltown show, the small Model 40 combine was also fitted with a 2-row corn attachment, the 205

E This August 1959 picture shows the 8′-cut 40 in its normal grain format. It was also available with a 10′ cut and had a 42-hp John Deere engine.

A Larger Combine Demanded

But the demand for a larger machine for the large farmer and custom operator grew as the 1950s progressed, so the company's answer was a 10″ wider, 4-straw-walker version of the 55 to be known as the 95. It was offered with all the options of its smaller counterpart. On a visit in 1959, the writer was able to purchase a 95 for use on his own farm, thus putting into retirement one of the previously mentioned 55 machines.

The more extensive updating of these self-propelled combines will be dealt with in the next book in this series, but before the end of the 50s the engines on all models had been changed to 4- and 6-cylinder John Deere, built in Dubuque, and the steering wheel was inclined like that of an automobile.

B The 95-R Rice combine is seen here in September 1957 in a heavy crop

A Deere's largest self-propelled combine in the fifties was to be the 95, seen here in September 1957 with draper pickup and straw spreader. The 95 was available with 12′ pickup platform or 12′-, 14′-, 16′- or 18′-platform. The John Deere 6-cylinder engine gave 80 hp, and the separator was 40″ wide with a 40″ × 22″ cylinder.

C An October 1958 photograph of a 95-R in a standing crop of rice. The 95-R Rice combines were equipped with 18.26 tires, as were the 55-Rs.

The following month a 95-RC on crawler tracks, and with the larger grain tank extension, harvests another good crop of rice

In August 1957, a 95-H Hillside model climbs safely around a western wheat field

Windrowers

In many parts of North America, farmers preferred to windrow their grain to avoid hail damage. It also meant that any green weeds or late-ripening crop could dry out before the combine, equipped with a windrow pickup, was brought into the operation.

Initially binders without their tying mechanisms were tried as windrowers. When combines arrived on the scene, their headers were adapted for use as windrowers, but soon specialized machines in varying sizes became the norm. Examples of both 9′ cut and 16′ cut are illustrated, driven from the tractor's power take-off.

As with combines, the demand arose for self-propelled versions, and in due course these too were added to the line, initially equipped with a small Wisconsin V-4 motor.

B An October 1929 picture of a 16′-cut No. 3 combine header windrowing a late crop of grain behind a Model "D" tractor

C The No. 3 combine header as used for windrowing

A Photograph of a Windrower pulled by an "M" tractor in 1950

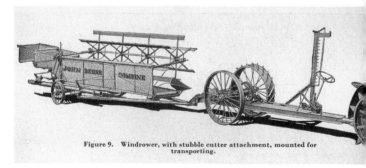

Figure 9. Windrower, with stubble cutter attachment, mounted for transporting.

D A similar unit arranged for transport behind a stubble cutter attachment

1. **Big Daily Capacity**—Lightweight, lighter-running John Deere 12- and 16-Foot Power-Driven Windrowers have big capacity of 50 to 60 acres per 10-hour day.

2. **Low-Cost Operation**—Rugged, quality construction assures more years of service with lower upkeep costs. Only 2-plow tractor power is required.

3. **Adjustable, Ground-Driven Reel**—A convenient lever within easy reach of the operator raises or lowers the reel. Reel is adjustable from 2 to 20 inches above the cutter bar.

4. **Wide Cutting Range**—The cutting height of the spring-balanced platform is easily set by means of handy crank adjustments on the main and grain wheels. The 3- to 26-inch cutting range is ample for all conditions. Platform is tilted by a handy lever within easy reach of the operator on the tractor seat. Powr-Trol attaching parts are available for effortless tilting of the platform.

5. **Heavy-Duty Platform Canvas**—The extra-heavy duck canvas assures long life and low upkeep.

6. **Efficient Windrow Layer**—Steel fingers gently ease cut material out on top of the stubble.

7. **Adjustable Windrow Widths**—Windrow widths are regulated to meet conditions by an adjustable, hinged, mold-board-type shield.

8. **Rugged Cutting Parts**—The same high-quality cutting parts as used in John Deere mowers and combines assure cleaner cutting and longer life.

9. **Durable Power Drive**—Power is delivered in a straight line from tractor to windrower. Power drive is carried on high-grade bearings. Forged steel universal joints provide ample flexibility.

10. **Smooth, V-Belt Drives**—Sickle and platform canvas are driven by sturdy V-belts, assuring smoother operation, greater flexibility, and longer life.

11. **Slip Clutch Protection**—Slip clutches on the power shaft, reel drive shaft, and platform canvas drive roller protect important units from breakage. High-pressure, grease-gun lubrication lengthens life and lowers upkeep costs.

12. **Special Equipment**—A shield for the grain side of the reel, a special sprocket to reduce reel speed, a light crop attachment for the 16-foot Windrower, and a rubber-tired transport truck are available as extra equipment.

CHECK
THESE OUTSTANDING
QUALITY FEATURES

(5)

E Power-driven 16′-cut binder windrower advertised in a December 1953 brochure

F Two Model "D"s with rear extension rims put windrows of grain side-by-side by using the two alternative methods of windrowing with 16′ headers from No. 3 combines in August 1930

Threshers

Although other manufacturers had been in the thresher business from the previous century, it was not until 1929 that the company acquired the ailing Wagner-Langemo Company of Minneapolis and its stock of threshing machines. These were marketed for that year and the next as the Grain-Saver thresher and were identified by the rear main straw blower being alongside rather than behind the machine.

In 1931, they were updated with the more conventional blower position and were known as the Light-Running model. Both types had a rack system of straw-grain separation, and it was not until 1937 that another series of two models with straw walkers was introduced.

The New Straw-Walker Models

The earlier models had extra width (compared with the competition), which was used as a bonus selling point, but with the introduction of walkers, they were reduced in width to the more normal 22″ × 36″ and 28″ × 42″. Later models were offered with a rubber-tire option, and after the war the machines were painted green and yellow, because of the difficulty of obtaining galvanized steel. The grain harvest was thus fully provided for by the company through the postwar years.

A A later type John Deere straw-walker thresher is resting after threshing at the Mount Pleasant, Iowa, show held in McMillan Park each Labor Day weekend. This occasion was in 1984.

B Another late straw-walker thresher, this time on rubber tires, on show at the Waterloo, Iowa, celebrations

C An August 1935 threshing scene, with a Light-Running thresher and some interesting methods of transporting both the unthreshed and threshed grain

Corn Pickers

Prototype corn pickers appeared from Harvester Works in 1926 at the same time as consideration was being given to a combine. Drawn models could be for horses or tractors. The No. 3 is illustrated as an early production model in 1927-28, followed shortly by the 10 and then the 15 one-row models. The first 2-row was the 20 in 1930, which was replaced in 1938 by the 21.

When tractors received lift mechanisms it became practical to mount corn pickers around the tricycle types, and the 2-row 25 and later 25-A were developed for this purpose.

B A 1927 studio photograph of one of the first Deere horse-drawn corn pickers with front tongue truck

A A rear view of a Model 10 corn picker with the rather unusual tank attachment for the cobs, towed by a "GP" tractor in November 1929

C This October 1928 photo shows a No. 3 corn picker behind a Model "C" tractor with horses and wagon in attendance for the picked corn

D By 1935, the tractor in charge of this 10 corn picker is one of the early Model "B"s with short frame and 4-stud fixing for the front column. They must have had dry ground to use tip-toe rear wheels.

E In November 1930, a "GP" tractor and 10 corn picker with a special hitch for towing the corn trailer at the same time

F A later Model "B" pulls its own wagon as well as the 10 picker.

A In July 1929, a Model "D," a 2-row 20 corn picker and a Deere wagon pose for the photographer

B The 25 in action on a Model "A" tractor in the 1936 corn harvest

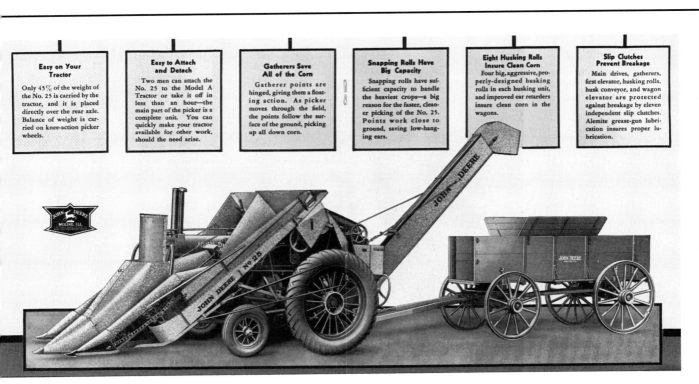

Easy on Your Tractor

Only 45% of the weight of the No. 25 is carried by the tractor, and it is placed directly over the rear axle. Balance of weight is carried on knee-action picker wheels.

Easy to Attach and Detach

Two men can attach the No. 25 to the Model A Tractor or take it off in less than an hour—the main part of the picker is a complete unit. You can quickly make your tractor available for other work, should the need arise.

Gatherers Save All of the Corn

Gatherer points are hinged, giving them a floating action. As picker moves through the field, the points follow the surface of the ground, picking up all down corn.

Snapping Rolls Have Big Capacity

Snapping rolls have sufficient capacity to handle the heaviest crops—a big reason for the faster, cleaner picking of the No. 25. Points work close to ground, saving low-hanging ears.

Eight Husking Rolls Insure Clean Corn

Four big, aggressive, properly-designed husking rolls in each husking unit, and improved ear retarders insure clean corn in the wagons.

Slip Clutches Prevent Breakage

Main drives, gatherers, first elevator, husking rolls, husk conveyor, and wagon elevator are protected against breakage by eleven independent slip clutches. Alemite grease-gun lubrication insures proper lubrication.

C Two men could attach this 25 corn picker to the "A" tractor in less than an hour, and less than half its weight was on the tractor as its front was carried on knee-action rubber-tired wheels

D This September 1940 photo shows the 21 corn picker, which replaced the 20, pulled by a styled "D" tractor. The old wagon has wooden wheels.

E By August 1939, the 25A corn picker had superseded the 25, and the "A" tractor was now styled

Semi-Integral Units Added

The 101 was a semi-integral one-row machine, shown in one picture on the preproduction "M" tractor in 1944, and this unit stayed in the line for a number of years.

The 25-A picker was replaced by the 226 first, and by the 227 near the end of our period covered, while the 101 gave way to the 127. Both burr mills and the 50 corn sheller were offered as alternative attachments behind the 227 in place of the standard trailer elevator.

B In November 1952, a 50 tractor provided the necessary power for this 101 picker

A In November 1944, a preproduction Model "M" tractor is the motive power required to operate this 101 wheel-and-drawbar-mounted picker in a heavy crop of corn. The 101 was of unique design and built especially for the "M" tractor.

C Experimental 40 XT-16 tractor with a 127 corn picker and 953 wagon picks 85-bushel corn on a Dubuque, Iowa, farm in October 1954

E November 1951 saw this 227 corn picker mounted on a late "A" tractor with a typical Midwest farmstead in the background

D A 1947 photograph of a 226 corn picker mounted on a tractor

F The following year one of the new 50 tractors acts as host to this 227 picker and John Deere wagon

A Corn Snapper

The Model 100 corn snapper was added to the line in 1950. This machine was identical to the 101 picker, except that the husking bed was replaced with an auger, allowing complete unhusked ears to be harvested.

B "MT" tractor pulling a 100 corn snapper

A This 227 corn picker mounted on a "30 series" tractor in October 1959 has a burr mill attachment in place of the normal elevator

C A preproduction 50 corn sheller (designated XO5-2) on a 227 corn picker with 720 tractor is seen in this September 1957 studio view

D By November 1958, the 50 sheller was in production. It is shown here on a 227 picker mounted on a 630 tractor

Cotton Harvesters

Both planting and cultivating of cotton as well as corn had been well covered by the company from early days, but harvesting was a different problem. Early cotton harvesters of differing types were built experimentally, and were odd-looking machines.

The Model 30 went into production in 1930, followed by the 31 in 1932, but it was the integral 15 and later 16 which really changed the stripper cotton harvest outlook.

Complete self-contained pickers mounted on 3-wheel tractors operated in reverse became popular with the one-row No. 1, followed by the 2-row No. 8, a true self-propelled machine.

B A completely different approach to harvesting cotton— the tractor has mechanical lift

A An experimental cotton stripper in 1929, mounted on a "GPWT" tractor with unusual rear wheels

C By 1930, the 30 cotton harvester was in production, as this picture from a sales brochure shows

JOHN DEERE
No. 31
COTTON HARVESTER

D In 1932, the 30 was replaced by the 31. Both are shown for 2-horse operation, ground driven, but could be tractor drawn if required

E A No. 8 picker shown harvesting cotton on a Mississippi plantation in 1950

F The last word in mounted 2-row cotton strippers in 1958 was the 16, here mated with a 520 tractor

D

E

F

Modern One- and 2-Row Cotton Pickers

The 22 was a one-row picker which replaced the No. 1, using all its many good features, and could be mounted on a single-front-wheel 530, 630 or 730 tractor or earlier model of similar size. It went into full production for the 1958 harvest.

A smaller version of the 22, the 11, was introduced in 1959, replacing 25 hand pickers and designed for the grower with as little as 40 acres. Mounted on a 430, 530 or 630 tractor or earlier equivalent, it would handle cotton as tall as 3'. Like all the later models it featured Air-Trol for cleaner picking.

A larger machine developed at the same time as the 22 was the 2-row 99 self-propelled model of advanced design, for doing the work of 80 hand pickers. The 99-H was a high-drum machine, the 99-L a low-drum type.

B A 1952 studio photograph of a No. 8 2-row cotton picker

A The experimental one-row No. 1 cotton picker shown in the mid-fifties

C This August 1958 picture shows the new one-row Model 22 cotton picker mounted on a 720-N tractor. This Air-Trol picker replaced the No. 1, had a 1200-pound basket and was available in high-drum or low-drum form

D Another picture of a similar outfit—22 picker and 720 tractor—taken the same month in 1958, shows the reverse method of operation and the shielded exhaust muffler in lowered position

Ⓐ The "little giant" of cotton pickers, the 11 was designed for smaller acreages and cotton up to 3′ high and is seen here on a 420 tractor. It was an Air-Trol picker, like the larger 22.

Ⓑ Another picture of a similar outfit shows the single-front-wheel 420

Ⓒ The 99-H high-drum picker had 20 rows of spindles, the 99-L low-drum 14 rows, picking from both sides of each row. The Air-Trol system kept suction away from the row, allowing trash to fall out and giving a cleaner sample of cotton. The 99s had a 77-hp John Deere gasoline or LP-gas engine.

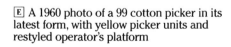 A three-quarter rear view of a 99 at work in December 1957

E A 1960 photo of a 99 cotton picker in its latest form, with yellow picker units and restyled operator's platform

Potato Harvesters

Early lifting of potatoes was accomplished by digging them out with a horse-drawn plow or lifter. The Hoover company developed a ground-driven machine with a chain conveyor bed, enabling a cleaning operation to be combined with the lifting process. In many parts of the world a "Hoover" indicated an elevating potato-lifting machine.

In due course these were available from Deere, first with an engine mounted overhead to power the conveyor but later with a power-driven shaft from the tractor. With this innovation it became possible to mount two elevator diggers on a common frame, thereby doubling the machine's capacity.

Deere's next development was to offer a Level-Bed model in both one- and 2-row sizes. The final phase saw the two beds merged into one wide bed, resulting in the Double Level-Bed model with really large capacity.

A derivative of the Double Level-Bed model was one with an elevating cross-conveyor at the rear of the machine, making it into a stone picker.

B This October 1938 picture shows the elevator digger with PTO drive from a Model "G"

A Taken from a 1927 sales brochure, this shows the early John Deere-Hoover ground-driven gear-drive potato digger with extension elevator and single-roller front truck

C This left-side view a year later shows a styled 4-speed "B" tractor pulling a rubber-tired PTO-driven digger at work

D A Double Level-Bed digger at work in September 1954 behind a 60-W wide-front-axle gasoline-engined tractor

E This June 1958 picture shows the 30 Double Level-Bed potato digger with stone picker attachment pulled by a 620-W wide-front-axle tractor

F Another view of the same outfit

Beet Harvesters

Sugarbeets were a much more difficult crop to mechanize for harvest. The tops were valuable as feed, so that a topper unit was required to remove them and windrow them separately. The beets then had to be gently lifted from the ground, and it was only then they could be elevated and cleaned to some extent, like potatoes.

The Model 54-A was usually mounted on an "A" tractor and could combine both these operations. One of these outfits, demonstrated quite extensively by the writer in the late '40s, was reasonably successful. A separate beet loader was usually used at that time, as the machine placed both the tops and beets in rows.

Later, first the 100 one-row harvester and subsequently the 200-A 2-row were included in the line for a number of years, with the 201 loader as its mate.

Dryers

Finally, in the harvesting area, dryers for grain, corn or other small seeds were introduced. The portable batch type represented by the 458 was popular at the end of the period we are covering, and both it and the 88 fan dryer, suitable for grain bins or hay, were demonstrated at the Marshalltown, Iowa, show.

A Beets and tops left in neat rows by a 54-A beet harvester mounted on a "BN" tractor in the mid-forties. The swinging rear boom allowed several rows of beets to be put together.

B The No. 6 beet loader, for use in collecting the beets from the above harvester; driven from the PTO of an "AN" tractor

C The 100 one-row beet harvester with 1½-ton cart, which could unload on the go or stationary. This gave the beet harvester the same advantages as a tanker combine in grain. The tractor is a 630-W and the date November 1958.

D The 458 offered push-button drying of shelled corn and small grain

E This 88 portable fan crop dryer is conditioning four trailer loads of hay with a 630 tractor supplying the power

Farmstead Equipment Review

This section covers a multitude of useful machines used often to keep the farm buildings and surrounding area tidy, to haul the crops grown, and to provide power handling of many farm chores.

The Rotary Cutter

The Gyramor rotary cutter was introduced to deal with brush and crops not suitable for sickle mower cutting, and has since developed into an extensive array of optional models.

We shall only cover the original model in this book, and two late additions, the 5'-cut 127 machine and the larger 5½'-cut 207 model. The multiplicity of types developed subsequently must await our next book.

The capacity of the rotary cutter was governed by the width of cut, but also by the depth of the body of the machine. All models were driven from the towing tractor's PTO, whether of drawn or integral type. The 127 had an 8"-deep body, and the 207 was 15" deep.

Manure Loaders

Manure loaders were first mounted at the rear of the tractor in 1939-40, and looked quite odd with their cable operation. Subsequently a more conventional front loader was developed for the tricycle-type tractors.

Ⓐ In June 1958, a 5'-cut 127 rotary cutter and 430 Row-Crop Utility tractor tackle tall brush

Ⓑ The deeper 207 Gyramor is mounted on a 630 tractor in November 1958

[C] Deere's first manure loader is seen here mounted on an "A" tractor. The team being loaded consists of a Model "H" spreader with an "H" tractor. The date of the scene is August 1939.

[D] A well-restored example of the same loader, seen at the Waterloo, Iowa, show on "A" No. 458,555 in July 1987

[E] Another early loader at a vintage gathering at Marshalltown, Iowa, in 1987, this time mounted on a "B" tractor

Additions to the Manure Loader Line

Various models of manure loaders were added to the line. The 25 fitted the "A" and "B" tractors; the 30 was produced for the "M" and continued in the line for its successors. The 40 covered the "MT" range, while the 45 was designed for the 50, 60 and 70 tractors and their following tricycle models. The 45-W covered the wide-axle row-crop tractors, while the 50, largest of the line, was for the "A," "B" and "G" models and their successors. The 35 was a late introduction for the 330, 430 and 435 tractors.

A This 1947 studio photo shows clearly the flywheel drive and cable operation of the loader and bucket tipping mechanism. The tractor is a late "B."

B A front-mounted 25, seen here on a "B" tractor, No. 4,266, at the Waterloo, Iowa, show in 1987

C A three-quarter rear view of the 25 loader

D With the arrival of the "M" series tractors, a new 30 loader was added to the line. It had two hydraulic cylinders for raising and lowering, but still had mechanical tipping of the bucket, and is shown here loading an "MT"-hauled "H" spreader, with an "MC" and dozer pushing the material to it.

E A December 1952 studio photo of the 40 loader on an "MT" tractor

F The 40 loader was designed for the "MT" tricycle tractor and its successors. It is seen here loading an "H" spreader hauled by a late "A" tractor

Ⓐ A 45 loader and "R" spreader with a brace of 620 tractors in September 1956

Ⓑ In May 1957, the 45 loader is on a 720 tractor, but the spreader is still a 95-bushel ground-driven "R"

C This 45-W loader on a 430RCU tractor makes an attractive scene with the typical Midwestern barn as backdrop

D Deere's largest loader in the fifties was this Model 50, shown mounted on the recently announced 50 tractor and loading a new "L" spreader in April 1952

Manure Spreaders

Manure spreaders came into the company's sales catalog with the acquisition of Kemp & Burpee and their Success spreader in the first decade of this century. The models "A," "B" and "E" followed—the "spreaders with the beater on the axle."

A nicely restored "B" spreader is shown at the Waterloo 150th year celebration, and another pulled by the experimental Melvin tractor about 1913.

B Deere's first spreader design, the Model "A," in original state on the Deere & Company display floor in 1987

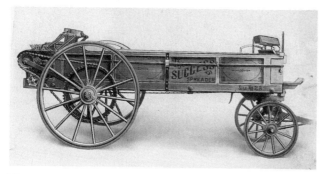

A A studio view of the Success spreader which Deere acquired from Kemp & Burpee when they purchased the company in 1910

C A Model "A" spreader behind the experimental Melvin tractor before World War I

D Still in use in October 1938, this "B" spreader shows only minor modifications from the "A," but still has the beater on the axle

E A beautifully restored "B" spreader belonging to Jim Hulse of Orangeville, Ontario, Canada, displayed at the Waterloo, Iowa, show in 1987

F This 1936 advertisement shows the Model "E" spreader fitted with an auger rear spreader

Power Drive and Rubber Tires

The first major development in spreaders was the provision of PTO drive for these essentially horse-drawn 4-wheel machines in the late twenties.

The rubber-tired "H" 2-wheel and "HH" 4-wheel were part steel, part wooden-bodied machines of modern appearance, and were succeeded in the line by five models of different capacities. Replacing the "H" and "HH" were the "L" and "M," respectively. The "R" was a 95-bushel ground-driven drawn machine, the "W" was 95-bushel and PTO driven, while the "N" was originally rated at 120 bushels, but was subsequently uprated to 134 bushels. The five models represented excellent coverage of the spreader market.

B A photo from *Better Farming 1949-50* of the "HH" 4-wheel spreader, horse-drawn in this case

A Pictured in March 1931, the Model "E" spreader modified for PTO operation

C May 1958 sees the Model "R" 95-bushel ground-driven spreader being hauled by a 530 tractor

D The "W" spreader was also 95-bushel capacity but was PTO driven, in this case from a 630 tractor. A 730 tractor and 45 loader do the loading.

E Deere's largest capacity spreader, the 134-bushel "N," was also PTO-driven. It is being loaded by a 620-W wide-axle tractor with 45-W loader.

Wagons, Buggies, and Surreys

A long line of wagons was offered by the company from late in the nineteenth century. Buggies and surreys had also been added through the Reliance connection.

A An early photo of a John Deere wagon equipped with a toolbox on its lower front endgate

B Illustration of a Deere buggy— "Deere vehicles are all right"

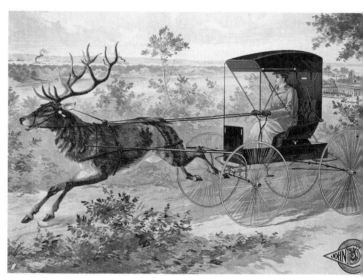

C An old advertising print of a John Deere buggy with a deer in the traces

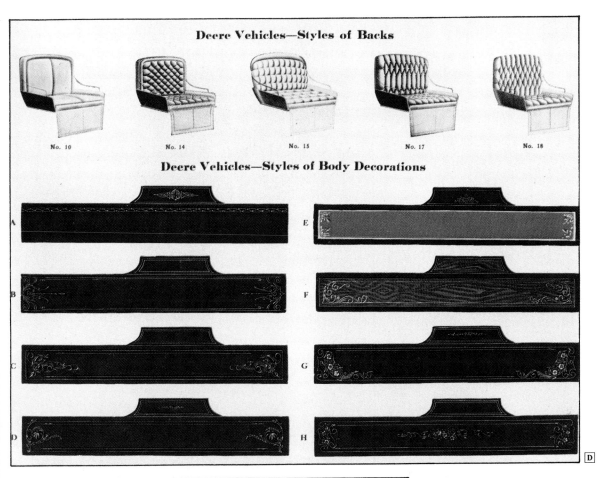

Deere Vehicles—Styles of Backs

No. 10   No. 14   No. 15   No. 17   No. 18

Deere Vehicles—Styles of Body Decorations

A   E

B   F

C   G

D   H

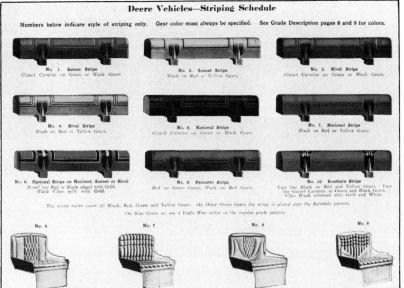

Deere Vehicles—Striping Schedule

Numbers below indicate style of striping only.   Gear color must always be specified.   See Grade Description pages 8 and 9 for colors.

No. 1. Sunset Stripe
Glazed Carmine on Green or Black Gears.

No. 2. Sunset Stripe
Black on Red or Yellow Gears.

No. 3. Rival Stripe
Glazed Carmine on Green or Black Gears.

No. 4. Rival Stripe
Black on Red or Yellow Gears.

No. 6. National Stripe
Glazed Carmine on Green or Black Gears.

No. 7. National Stripe
Black on Red or Yellow Gears.

No. 8. Optional Stripe on National, Sunset or Rival
Broad line Red or Black edged with Gold.
Black Clips split with Gold.

No. 9. Palmetto Stripe
Red on Green Gears, Black on Red Gears.

No. 10. Southern Stripe
Two line Black on Red and Yellow Gears. Two
line Glazed Carmine on Green and Black Gears.
Clips Black trimmed with Gold and White.

The above styles cover all Black, Red, Green and Yellow Gears. On Olive Green Gears the stripe is glazed over the regular Schedule pattern.
On Blue Gears we use a Light Blue stripe of the regular grade pattern.

No. 6   No. 7   No. 8   No. 9

D Different styles of body decorations and seat backs for Deere vehicles

E Striping schedule for Deere vehicles and four other seat styles

Wagons

In later times the practice was to offer a 4-wheel chassis on steel wheels or rubber tires, and let the customer choose bodies to his requirement. First the Big 3 consisting of the 943 Economy, 953 Standard and 963 Heavy-Duty models, and later the Big 4 with the addition of the 1064 Wide-Tread model, were listed.

Specialized wagons like the 15 and 16 tipping units and the small 2-wheel trailers for use with the farmer's car—the 20 and later the 25—expanded the line.

A The 963 gear with hydraulic 150 bolster hoist on a tipping wagon at the 1959 Marshalltown John Deere demonstration in Iowa

QUALITY FEATURES

- Ruggedly Built.
- Large, Strong Tilting Platform, Legal Width.
- 3-Ton Capacity.
- Dual Wheels with Timken Tapered-Roller Bearings.
- Low to the Ground, Easy to Load and Unload.
- Straight-Trailing, Light-Running.
- Powerful, Variable-Speed Loading Winch.
- Automatic Brakes Regular Equipment.

B The 3-ton 16 tipping wagon with built-in variable-speed loading winch, dual wheels and automatic brakes, pictured in a 1948 leaflet

C A factory picture of a 2-wheel 25 trailer in June 1936

D A familiar sight around Moline in the thirties and still remembered by acquaintances of the author, this coal wagon was built on the John Deere 802 all-steel gear

Your Choice of FOUR Modern, Precision-Built, Light-Running Gears

953 STANDARD
Today's top value in general-purpose gears, the 953 Standard permits highway speeds with payloads limited only by size and condition of the rubber tires. Precision-built for long, hard service, it deservedly is the No. 1 choice of thousands of value-wise farmers.

963 HEAVY-DUTY
Above you see the 963 Heavy-Duty equipped with 4-wheel hydraulic brakes (optional; must be factory-installed), 10-leaf spring bolsters, and double-acting coil stabilizer springs. The use of 8-ply 7.50x16 rib implement tires is recommended for carrying maximum loads.

943 ECONOMY
This wagon thoroughly fills the bill for lighter loads and low-cost dependability. It is economically but not "cheaply" built—cost has been reduced by such measures as eliminating bolsters and mounting the stakes directly on the heavy I-beam axles.

1064 WIDE-TREAD
The 1064 has 72-inch tread, compared with 62 for other John Deere Wagons . . . 80-inch over-all width, also wider than the others . . . and has the unusually tight turning radius of approximately 9 feet, 4 inches, with reach in short position, making it ideal for use behind corn pickers, field choppers, and balers.

E A May 1958 advertisement for Big Four wagon gear

Elevators

When Deere acquired the manure spreader from Kemp & Burpee, an elevator line was included. Elevators were originally operated with horse-power, but the tractor took the horses' role with a drive from its belt pulley and later its PTO. Variants appeared in the barn, in tubular form externally, and eventually in 18½″-wide elevators which could move bales as well. Combined with these elevators an overhead track for dropping small bales at random in the barn completed a one-man baling system.

B One tractor with a John Deere wagon empties its load of corn cobs into a portable grain elevator PTO-driven from a Model "B"

A A June 1927 photograph of a portable grain elevator, belt driven from a Model "D" tractor

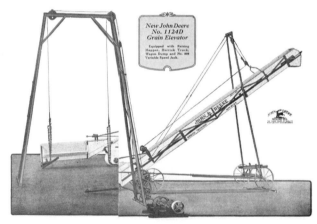

C Advertising material from a 1929 brochure for the new 1124D grain elevator equipped with raising hopper, derrick truck, wagon dump and 809 variable-speed jack

D The 18½"-width bale-size elevator with electric motor drive, connected to a barn bale conveyor as part of a one-man baling system

E Tubular portable steel elevator loading into a rail car at 15 to 20 bushels of grain a minute

F A November 1950 picture of an LUC motor-driven portable elevator unloading corn cobs from a John Deere wagon

Shellers

John Deere shellers started life as small hand-operated models, the one-hole No. 1, 1-A and 1-B, and gradually larger and more complicated units from the Nos. 3, 4 series, 6, 7, 9 and 10 up to modern models like the 71 and 43 toward the end of our timescale.

B The later No. 1-B corn sheller doing a clean job in May 1939

A No. 1 corn sheller on the Deere & Company display floor at Moline in July 1987. This unit is part of Deere's historical collection

C A steel-wheeled Case SC3 tractor is the motive power for this No. 7 corn sheller in March 1943, during World War II

D This 1933 wintry scene with the corn wagon on snow skids shows the No. 9 steel-cylinder sheller at work

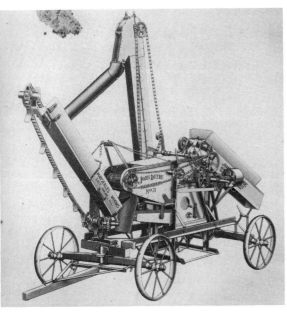

E A 1920 studio picture of the No. 3 corn sheller

F A 520 tractor and 43 corn sheller in a mid-1950s studio picture

Mills

In addition to shellers, hammer mills for grinding feed were added to the line in 6″, 10″ and 14″ sizes, with roughage mills, similar but with different feed, in the two larger sizes.

Ⓐ A belt-driven 6″ No. 6 hammer mill grinding corn cobs

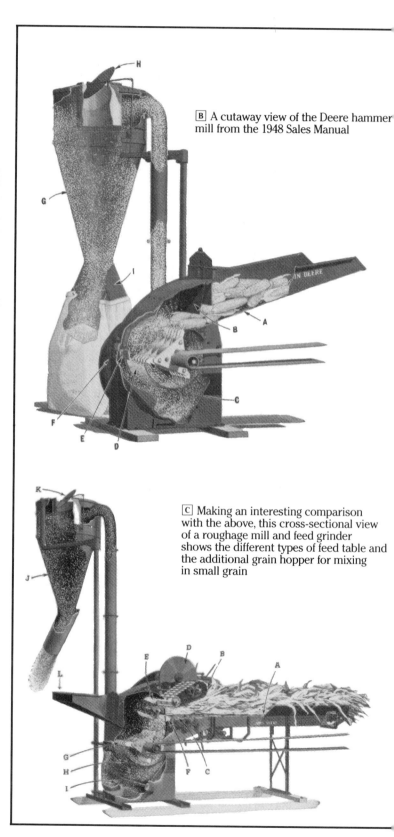

Ⓑ A cutaway view of the Deere hammer mill from the 1948 Sales Manual

Ⓒ Making an interesting comparison with the above, this cross-sectional view of a roughage mill and feed grinder shows the different types of feed table and the additional grain hopper for mixing in small grain

D This July 1952 picture of a 10″ 10A hammer mill shows it with the PTO-drive option. It also has retractable transport wheels.

Snow Plows

Snow plows, 60″ wide to fit the "M" tractor, a 72″ model for the "MT" and its successors, and a 90″ version for the "A," "B" and "G" and later similar-sized tractors, were useful additions in the snow areas of the central and northern states.

B A 420 industrial model fitted with a Welland-built snow plow, pictured in January 1957

A A preproduction Model "M" tractor, the XM12, with Touch-O-Matic control to raise and lower the integral snow plow, at Syracuse, New York, in July 1946

C An ABG-90 snow plow fitted to a 50 tractor in March 1953

D An experimental "MC" crawler tractor with 4-roller track frame and inside-frame hydraulic dozer. This unit preceded the 4- and 5-roller 40C models which followed the "MC" in 1953.

E This 420 industrial model has a 3-point linkage-mounted 80 rear blade in this September 1956 picture

Dozers, Blades, Scrapers and Earth Scoops

Dozers, blades, scrapers and scoops were a natural addition to the line, all made more readily available with the extension of hydraulics on all sizes of tractor.

For quite a time the Handy concrete mixer was another useful item to own about the yard.

Land Levelers and Landshapers

These implements had been included in the line from the purchase of Killefer and Lindeman, respectively. There was little in earthmoving equipment the farmer could not obtain from the company's dealers if he so wished.

B These two Model "DI" tractors work a site in July 1936. Both have twin rear wheels and front wheel weights and the nearer one is pulling a rolling scraper.

A In November 1952 a 40T tractor is at work with a No. 1 utility tool carrier fitted with a scraper attachment

C This 420U utility tractor is working with an integral rear-mounted 20 scoop in September 1956

Handy Farm Mixer and Engine
Mounted on Portable Truck

D

D A 1929 sales literature illustration
of the Handy concrete mixer fitted with an
"E"-type engine drive

E A Model "M" tractor pulls a John Deere-
Lindeman LS200 landshaper and a Model "A"
follows with an LS500 near the Lindeman
Works at Yakima, Washington, in July 1948

E

Stationary Engines

One final range of machines must be mentioned to complete this review. Before purchasing the Waterloo Gasoline Engine Co., Deere had been marketing the Root and Vandervoort line of engines.

Even earlier than that they had sold the New-Way engines for some applications. With the Waterloo purchase they acquired a line of seven engine sizes, from 2- to 14-hp Model "H"s to the 25-hp Model "T."

B A unique engine, No. 83 of the Waterloo Gasoline Traction Engine Co. of Waterloo, Iowa. The company was only in business under this title from 1893 to 1895. The engine belongs to Charles Wendell of Iowa, who gave the author this photograph.

A A 5-hp Waterloo Boy Model "H," No. 232,145

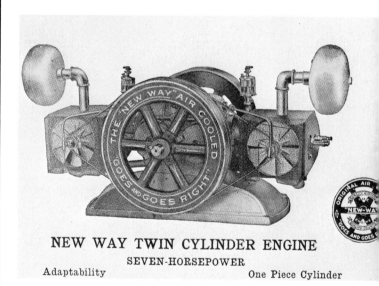

NEW WAY TWIN CYLINDER ENGINE
SEVEN-HORSEPOWER
Adaptability   One Piece Cylinder

C A New Way 7-hp 2-cylinder air-cooled horizontally opposed stationary engine, used by the company on occasion before the association with Root & Vandervoort

D Nicely restored 2-hp Root & Vandervoort engine No. R219,187 at the Mount Pleasant, Iowa, show on August 30, 1984

E A 2-hp Waterloo Boy "H" stationary engine, No. 232,575, at the Mount Pleasant, Iowa, show in 1984

F This Waterloo Boy "H" was on show at Stockton, Kansas, on September 27, 1986

Deere Updates

In 1923 the "E" series of 1½-hp, 3-hp and 6-hp engines replaced the six "H" models, and the "W" 27-hp unit—a stationary version of the "D" tractor—replaced the "T." The first "W"s had the spoke flywheel like the tractor, but in time this became the solid type.

[A] A 1½-hp "E" with jack pump forms a neat outfit at the Waterloo show. Its serial number is 318,254 and its proud owner is Robert Dufel of Hudson, Iowa.

[B] A cutaway picture in 1926 of the type "E" stationary engine which had replaced the Waterloo Boy "H" series in 1923

John Deere Type EK Kerosene Engine

JOHN DEERE

The Enclosed Engine That Oils Itself

The John Deere type EK kerosene engine will give real satisfaction on any job within its power range. It will produce reliable and economical power, month after month, year after year with minimum delays and repair expense.

Operates on Four-Stroke Principle

The John Deere Kerosene Engine operating on the four-stroke principle and controlled by a throttle-governor, delivers a power stroke every revolution of the crankshaft. This insures a steady, even flow of power, and maintains an even temperature for the successful burning of kerosene and other fuels of similar grade.

Simple Fuel System

In the simple fuel system there are no cams, pumps, fuel return lines or carburetor float.

From the tank, protected in the base of engine, the fuel is drawn to the carburetor by suction, as required. The throttle valve in carburetor is connected to the governor and regulates the fuel to the load. A speed-adjusting screw permits the speed of engine to be varied at the will of the operator. The small bowl on carburetor is used for gasoline when starting the engine.

All Vital Parts Completely Enclosed

All the important parts—crankshaft bearings, connecting rod bearings, cylinder, governor, timing gears, etc.—are completely protected within a dust-proof housing.

There is no opening for dust, sand or other foreign matter to get into the vital parts of the John Deere.

[C] This 1929 sales brochure page describes the kerosene-fueled version of the "E," the Model "EK"

D A rare photograph of five Model "E" engines awaiting export all over the world, outside S.A. Asquith & Bro., probably in Chicago. The engines are two 1½-hp, two 3-hp and a 6-hp.

E Lloyd Bellin's Waterloo Boy Model "T" 25-hp stationary engine No. 1,041 at home near Isanti, Minnesota, complete with cart and driver's seat. These later Model "T" engines had a series of serial numbers independent of the tractor equivalent. The collection of pedal tractors makes an interesting backdrop.

F Another Model "T," No. 6,542. This one was on show at Waterloo in 1987 and is the earlier type with dual gasoline-kerosene fuel tank, numbered among the "R" tractors. The owner is Travis Jorde of Rochester, Minnesota, who at the time of writing also owned the earliest known Waterloo Boy tractor.

A The Layhers of Wood River, Nebraska, own this Type "T" stationary tank-cooled 24-hp engine with friction clutch pulley, mounted on a steel frame. No. 1,229

B Another view of the same engine, alongside a spoke-flywheel Model "W," the type which superseded it

C The 35-hp Model "W" shown above, No. 235,549

D Another view of No. 235,549 showing the channel-steel mounting frame

E A three-quarter rear view of W-111 35-hp portable engine No. 235,539 with spoke flywheel and mounted on a steel truck; the property of the Bellins of Isanti, Minnesota

F A Model "D" tractor converted to a logging winch working in the woods in September 1937. The front axle support bracket, rear transmission case and back axle show clearly in this view.

The End of an Era.

And so this broad survey is complete. Bicycles were featured briefly in the 1890s and even a car, the Deere-Clark, was offered in 1906 but disappeared as quickly as it had arrived.

The story began with an earnest young blacksmith producing a self-cleaning steel plow, an innovation that aroused wonder and solved a difficult problem. In 1959, the present story ends with a huge 8-bottom fully integral plow sitting in a marquee at the Dallas dealers' presentation of new machines, and everyone wondering what massive tractor could pull it, let alone lift it.

B An 1890s John Deere bicycle displayed at the Deere & Company Administrative Center in July 1987 as part of the sesquicentennial celebrations

Deere Bicycle Trade Mark.

A This Deere bicycle trademark appeared in the August 1896 issue of the *Farm Implement News*

C A sales leaflet from 1907 of the Deere-Clark Type "B" Motor Car

1907 TYPE "B"

4 CYLINDER   25-30 H. P.   4 PASSENGER

PRICE $2500

SPECIFICATIONS

Deere-Clark Motor Car Co.
112 Black Hawk Avenue, Moline, Illinois

D The shape of things to come—a fully mounted 8-furrow plow appeared in the summer of 1959......

PART III

APPENDICES

Model	Years	Serial Numbers	Features & Notes
Froehlich Prototype	1892	One built	Vertical
Froehlich	1893	Four built	Single cylinder, 25 hp.
Waterloo Gasoline Engine Co.	1896	One built	Horizontal
Waterloo Gasoline Engine Co.	1897	One built	Single cylinder, 25 hp.
Waterloo Boy Standard Model "TP"	1912-1913	Numbers built	4-cyl. 5½″ × 6″, 650 rpm. Cross-Mounted Engine
W.B. Sure Grip, Never Slip	1913	not recorded.	Rear crawler tracks with same engine.
W.B. Light or "L"	1913-1914	1000-1025 1034 1044 1253	2-cyl. opposed, 15 hp. 5½″ × 7″
W.B. "C"	1913	No record of numbers built.	2-cyl. opposed, 15 hp. 5½″ × 6″ 4-wheel drive.
W.B. "R" Style A	1914	1026-1033 1035-1043 1045	2-cyl. horizontal twin 24 hp. 5½″ × 7″ integral head and block
W.B. "R" Style B	1914-1915	1046-1070 1072-1075	Same engine as style "A."
W.B. "R" Style C	1915	1071 1076 1077	Same engine as style "A."
W.B. "R" Style D	1915	1078-1143	Same engine as style "A."
W.B. "R" Style E	1915	1144-1252 1254-1316 1601 1606/8/10/13 1619-1653	2 cyl. horizontal twin 25 hp. with integral head & block 6″ × 7″
W.B. "R" Style F	1915-1916	1317-1509	No. 1407 first tractor shipped to U.K. Same engine as style "E."
W.B. "R" Style G	1916	1510-1600/2-5 1607/9/11/12 1614-1618	Same engine as style "E."
W.B. "R" Style H	1916	1654-3130 3132-3208/10-14 3216-3246/48-81 3283-3297 3299-3307/9-16 3318-3330/2 3334-3342/4-3365 3368-71/3-5/7-83 3385-90/93-3399 3402-20/4-9/31/4 3435/7/40/1/3-5 3447-61/3/4/6-82 3484-3502/4-10 3512-5/7/9-21/3 3525/6/8-32/4 3536-3545/50-54 3557/9/60/2-5 3572-74 3593/6/8	2-cyl. horizontal twin 25 hp. with separate head & block 6″ × 7″ No. 1747 oldest known tractor at present in U.K. No. 3598 shipped July 21, 1916
W.B. "R" Style I	1916	3131 3209 3215 3247 3282 3298 3308/17/31/33/43 3366/67/72/76/84 3391/92 3400/1 3421-23/30/2/3/6 3438/9/42/6/62/5 3483 3503/11/6/8 3522/4/7/33/5 3546-9/55/6/8/61 3566-71/5-92 3594/5/7/9-3609 3611-4199 4204/5 4208/11/34/5/68 4270/6/8 4279	All styles of Model "L," "R," and "N" have serial numbers inter-mixed with other styles and models, including the "T" stationary engines. No. 4279 shipped Oct. 25, 1916
W.B. "R" Style K	1916-1917	4200-3/6/9/10/2/3 4215-24/6-30/2 4236-43/5/7-51/4 4269/72/3/5/81/3 4285/7 4302 4791/7 4800/27/41	

JOHN DEERE TRACTOR MODEL PRODUCTION YEARS

Model	Year Start	Year Finish	Serial Number First	Serial Number Last	Features & Notes
B. "R-M"	1917	1918	6,728	10,336	Single-speed, 6½″ × 7″ engine.
B. "N"	1917	1924	8,378 31,320	30,400 31,412	2-speed, 6½″ × 7″ engine.
WD	1918	1919	1918-01	1918-100	All-wheel-drive 'Dain.'
	1924	1924	30,401	30,450	First 50 fabricated front axle, 26″ spoke flywheel, 6.5″ × 7″ engine, L.H. steer.
	1924	1924	30,451	31,279	26″ spoke flywheel, 2-speed, ladder-sided radiator.
	1924	1926	31,280 31,413	31,319 36,248	24″ spoke flywheel, 2-speed.
	1926	1927	36,249	53,387	2-speed, keyed solid flywheel.
	1927 1927	1930 1940	53,388 53,388	109,943 149,187	2-speed, splined flywheel, 6.75″ × 7″ engine. Industrial model.
	1930	1934	109,944	119,099	R.H. worm steer.
	1935	1938	119,945	143,799	3-speed, cast front.
	1939	1953	143,800	191,578	Styled model. Electrics optional.
	1953	1953	191,579	191,670	191,670 shipped Mar 18, '54 "Streeter" Models.
	1927		1	4	No. 1 tricycle type.
	1927		9		No. 9 tricycle type.
	1927		101	125	Except No. 123. 16 rebuilt and renumbered.
	1927		200,001	200,069	Except, 200,009/10/24.
	1927		200,080		
	1927	1928	200,099	200,110	Except, 200,100/1/4/5. 37 rebuilt and renumbered.
	1928	1930	200,211	223,802	5.75″ × 6″ engine.
	1930	1935	223,803	230,745	6″ × 6″ engine.
Tricycle	1928	1929	200,264	204,213	See Appendix A-8 for 23 serial numbers.
	1931	1935	15,000	15,732	Orchard model.
WT	1929	1930	400,000	402,039	5.75″ × 6″ engine, side steer.
	1930	1932	402,040	404,809	6″ × 6″ engine, side steer.
	1932	1933	404,810	405,254	6″ × 6″ engine, overhead steer.
	1930	1930	5,000	5,202	5.75″ × 6″ engine. 68″ rear tread.
& 3	1933		410,000	410,007	See Appendix A-9.
	1934	1935	410,008	414,808	5½″ × 6½″ engine, open fan shaft.
N, AW	1935	1938	414,809	476,999	Unstyled. AN single front wheel & AW wide axle.
, AWH	1937	1938	469,668	476,999	Unstyled Hi-Crop model.
	1938	1940	477,000	498,999	Styled 4-speed. AN, AW, ANH, AWH included.
	1941	1947	499,000	583,999	Styled 6-speed. Options as above. 5.5″ × 6.75″ engine.
	1947	1952	584,000	703,383	Late model with electric start, pressed steel frames.
	1951	1952	665,000	703,383	Hi-Crop version.
	1935 1936	1940 1941	250,000	259,999	Standard 4-wheel. 5½″ × 6½″ engine. Offset radiator cap.
	1941	1949	260,000	271,999	Standard 4-wheel. 5½″ × 6.75″ engine, center radiator cap.
	1935	1936	250,000	252,723	Orchard model.
	1936	1940	1,000 1,525 1,697 1,800 1,850	1,497 1,693 1,771 1,836 1,891	Streamlined model.
	1940	1949	260,000	271,999	5.5″ × 6.75″ engine, center radiator cap.
AO	1949	1952	272,000	284,074	Styled, electric start.

WATERLOO BOY TRACTOR PRODUCTION YEARS (Cont'd)

Model	Years	Serial Numbers	Features & Notes
W.B. "R" Style L	1916-1917	3610 4207/14/25 4231/3/44/6/52/3 4255-67/71/4/7/80 4282/4/6/8-4301 4303-6727 6730/1/3/5 6774 6781-4/7/8 6869 6870/2-7/80/1/4-6 6888-90 6900/17/8 6979 7127	
W.B. "R" Style **M**	1917-1918	6728/9/32/4/7/9 6742-73/5-80/5/6 6789-6868/71/8/9 6882/3/7/91-9 6901-16/19-6978 6980-7126/8-64 7166/7/9-81/83 7185-7919/21-8050 8052-8377/9/81 8384-8408/10-8606 8608-11/3/4 8616-8723/5-33 8735-7/9/40/3-78 8808-40/2/4-51 8853-8924/39/40 8942-6/7/9-82 8984-9012/4/5/7-9 9021-45/8-53 9056-9131/5/8 9149 9151/3-60/2/6 9171-97/99-9213 9231/40/4-7/50-3 9232-39/41-3/54-6 9901 10000-19 10138/91 10200 10218/27/81/7/90 10292/8 10300-2 10304/6/7/10/11 10313/7/9-21/6/7 10329/31-3/5/6	6½″ × 7″ motor. 12th and last style of Model "R."
W.B. "N" 2-speed	1917-1924	8378/80/2/3 . . . . 8409 8607/12/5 8779-8807 8925-8937/45/48 8983 9013/6/46/7 9127-9/32/3/9 9140-48/50/2/61 9163-65/67-70 9214-6/9-30 9248 9249/57-61 9324 9456 9558 9901 10020-10137 10139-10190/2-9 10201-17/19-26 10228-10280/2-6 10288/9/91/3-7/9 10303/5/8/9/12 10314-16/8/22-25 10328/30/34/37 10344-30400 31320-31412	First four model "N" 2-speed tractors all sold to P.J. Downes, Minneapolis, MN. Auto steering 20834 on. Riveted frame 28094 on. Numbered among the Model Ds and built in 1924.
W.B. "T" Portable	1917-1925	7165/8/82/4 7920 8051 8724/34/8 8741/2 8841/3/52 8938/41 9020 9054/5 9134/6/7 9198 9217 9218 9262-9287 1001-1338	Built April 30, 1920—July 7, 1925

Serial Number	Type	Destination
1000	3-wheel center	Northey, California.
1001	3-wheel R.H. offset	Cleveland, California.
1002	4-wheel	Northey, California.
1003	4-wheel	San Francisco, California.
1004	4-wheel	Buhl, Idaho.
1005	4-wheel	Cleveland, Calif.
1006	4-wheel	Wayland, Mo.
1007	4-wheel	Bristol, Ill.
1008	4-wheel	Grover, Colo.
1009	4-wheel	Storm Lake, Iowa.
1010	4-wheel	Lakeville, Minn.
1011	3-wheel	Cleveland, Calif.
1012	4-wheel	Spencerville, Ohio.
1013	4-wheel	Minneapolis, Minn. (P.J. Downes)
1014	4-wheel	Goshen, Ind.
1015	4-wheel	Ryan
1016	4-wheel	J.F. Search
1017	4-wheel	Palamet, Ill.
1018	4-wheel	Earlville, Ill.
1019	3-wheel	Cleveland, Calif.
1020	3-wheel	Cleveland, Calif.
1021	3-wheel	Cleveland, Calif.
1022	3-wheel	Cleveland, Calif.
1023	3-wheel	Cleveland, Calif.
1024	3-wheel	Cleveland, Calif.
1025	4-wheel	Valparaiso, Ind.
1034	4-wheel	Minneapolis, Minn. (P.J. Downes)
1044	4-wheel	Rock Valley, Iowa.
1253	4-wheel	Minneapolis, Minn. (P.J. Downes)

Test No.	Date	Model	Fuel	Motor Rpm	Motor Size (in.)	Horsepower Brake	Horsepower Drawbar	Max. Lbs. Pull	Shipping Weight
1	3/31/20	WB N	K	750	6.5 × 7	25.00	15.98	2900	6183
102	4/11/24	D	K	800	6.5 × 7	30.40	22.53	3277	
146	10/24/27	15-27	K	800	6.75 × 7	36.98	28.53	4462	4917
153	10/22/28	GP	Gas	900	5.75 × 6	24.97	17.24	2489	4265
190	5/ 4/31	GP	Dist.	950	6 × 6	25.36	18.86	2853	4925
222	4/19/34	A	Dist.	975	5.5 × 6.5	24.71	18.72	2923	4059
232	11/15/34	B	Dist.	1150	4.25 × 5.25	16.01	11.84	1728	3275
236	6/26/35	D	Dist.	900	6.75 × 7	41.59	30.74	4037	5690
295	11/15/37	G	Dist.	975	6.12 × 7	35.91	27.63	4085	5160
305	9/ 6/38	B	Dist.	1150	4.5 × 5.5	18.53	14.03	2088	3390
305	9/ 6/38	B	Dist.	1150	4.5 × 5.5	18.53	16.44	2690	4360
312	10/31/38	H	Dist.	1400	3.56 × 5	14.84	12.48	1839	3035
313	11/ 4/38	L	Gas	1550	3.25 × 4	10.42	9.06	1235	2180
335	11/13/39	A	Dist.	975	5.5 × 6.75	29.59	26.20	4110	6410
350	7/22/40	D	Dist.	900	6.75 × 7	42.05	38.02	4830	8125
366	11/ 7/40	B	Dist.	1150	4.5 × 5.5	20.52	18.26	2756	4545
373	6/20/41	LA	Gas	1850	3.5 × 4	14.34	13.10	1936	3490
378	10/27/41	AR	Dist.	975	5.5 × 6.75	30.33	26.52	4248	6350
380	4/28/47	B	Gas	1250	4.69 × 5.5	27.58	24.62	3437	5178
381	5/ 9/47	B	Dist.	1250	4.69 × 5.5	23.53	21.14	3689	4996
383	6/ 5/47	G	Dist.	975	6.12 × 7	38.10	34.49	4394	7442
384	6/ 7/47	A	Gas	975	5.5 × 6.75	38.02	34.14	4045	6574
387	10/ 6/47	M	Gas	1650	4 × 4	20.45	18.15	2329	3952
406	4/19/49	R	Dies.	1000	5.75 × 8	51.0	45.7	6644	10398
423	9/ 7/49	MT	Gas	1650	4 × 4	21.6	18.8	2385	3929
429	10/11/49	AR	Gas	975	5.5 × 6.75	39.1	34.9	4431	7367
448	7/20/50	MC	Gas	1650	4 × 4	22.2	18.3	4226	4293
472	5/26/52	60	Gas	975	5.5 × 6.75	41.6	36.9	4372	7413
486	10/15/52	50	Gas	1250	4.69 × 5.5	31.0	27.5	3504	5439
490	4/15/53	60	T.F.	975	5.5 × 6.75	33.3	30.1	4499	6772
493	5/15/53	70	Gas	975	5.87 × 7	50.4	44.2	5453	8677
503	9/ 9/53	40	Gas	1850	4 × 4	25.2	22.9	3022	4553
504	9/ 9/53	40S	Gas	1850	4 × 4	24.9	22.4	2543	4189
505	9/ 9/53	40C	Gas	1850	4 × 4	25.0	20.1	4515	4669
506	9/25/53	70	T.F.	975	6.12 × 7	45.0	41.0	5102	8135
507	9/25/53	50	T.F.	1250	4.69 × 5.5	25.8	23.2	3583	5521
513	11/ 6/53	60	L.P.	975	5.5 × 6.75	42.2	38.1	5352	7609
514	11/ 6/53	70	L.P.	975	5.87 × 7	52.0	46.1	6127	8768
528	10/19/53	70	Dies.	975	6.12 × 6.37	51.5	45.7	6189	9028
540	5/ 9/55	50	L.P.	1250	4.69 × 5.5	32.3	29.2	3466	5645
546	6/ 3/55	40S	T.F.	1850	4 × 4	20.9	19.0	2511	4159
567	10/27/55	80	Dies.	1125	6.12 × 8	67.6	61.8	7394	11485
590	9/ 5/56	520	L.P.	1325	4.69 × 5.5	38.09	34.17	4659	6605
591	10/ 6/56	620	L.P.	1125	5.5 × 6.37	50.34	45.78	5920	8769
592	9/13/56	520	T.F.	1325	4.69 × 5.5	26.61	24.77	4660	6430
593	9/17/56	720	L.P.	1125	6 × 6.37	59.61	54.17	6697	9055
594	9/18/56	720	Dies.	1125	6.12 × 6.37	58.84	53.66	6547	9241
597	10/ 6/56	520	Gas	1325	4.69 × 5.5	38.58	34.31	4723	6505
598	10/10/56	620	Gas	1125	5.5 × 6.37	48.68	44.16	6122	8655
599	10/13/56	420W	Gas	1850	4.25 × 4	29.21	27.08	3790	5781
600	10/13/56	420S	T.F.	1850	4.25 × 4	23.47	21.89	2734	4311
601	10/15/56	420C	Gas	1850	4.25 × 4	29.72	24.12	4862	5079
604	11/ 1/56	620	T.F.	1125	5.5 × 6.37	35.68	32.66	6107	8505
605	11/ 6/56	720	Gas	1125	6 × 6.37	59.12	53.05	6647	8945
606	11/12/56	720	T.F.	1125	6 × 6.37	45.33	41.29	6608	8959
632	10/14/57	820	Dies.	1125	6.12 × 8	75.60	69.66	8667	11995

P.T.O.

Test No.	Date	Model	Fuel	Motor Rpm	Motor Size (in.)	Horsepower Brake	Horsepower Drawbar	Max. Lbs. Pull	Shipping Weight
716	9/ 8/59	435	Dies.	1850	3.87 × 4.5	32.91	28.41	4241	6057
717	9/ 9/59	440	Dies.	1850	3.87 × 4.5	32.70	27.51	4362	6375
718	9/10/59	440	Gas	2000	4.25 × 4	31.06	26.90	3950	5975
719	9/10/59	440C	Dies.	1850	3.87 × 4.5	32.88	26.15	7060	7281
720	9/10/59	440C	Gas	2000	4.25 × 4	31.91	24.23	6548	6919

Model	Year Start	Year Finish	Serial Number First	Serial Number Last	Features & Notes
B, BN, BW	1935	1937	1,000	42,199	4.25″ × 5.25″ engine, short frame.
B, BN, BW	1937	1938	42,200	59,999	4.25″ × 5.25″ engine, long frame. BN single wheel and BW wide front axle.
BNH, BWH	1937	1938	46,175	59,999	High-clearance models.
B, BN, BW, BNH, BWH	1938	1940	60,000	95,999	Styled 4-speed, 4½″ × 5½″ engine.
B	1941	1947	96,000	200,999	Styled 6-speed. BN, BW, BNH and BWH options as above.
B	1947	1952	201,000	310,775	Electric start, pressed steel frames. 4.69″ × 5.5″ engine.
BR & BO	1935	1938	325,000	328,999	4.25″ × 5.25″ engine, no decompression taps
BR & BO	1938	1947	329,000	337,514	4.5″ × 5.5″ engine.
BI	1936	1941	325,617	332,157	Industrial version.
BO-Lindeman	1939	1947	329,000	337,514	Crawler tracks fitted by Lindeman, Washington.
G	1938	1941	1,000 / 7,100	7,000 / 12,192	6.125″ × 7″ engine. Largest row-crop tractor. 4-speed.
GM	1941	1946	13,000	22,112	Styled model, hand start, single front wheel and wide axle option, 6-speed.
G, GN, GW / GH	1947 / 1951	1953 / 1953	23,000	64,530	Electric start. / Hi-Crop.
H, HN	1939	1947	1,000	61,116	3-speed, optional single front wheel
HWH / HNH	1941 / 1941	1942 / 1942	28,493 / 30,172	42,842 / 42,726	High clearance for California.
Y	1936	1936	NOT NUMBERED		26 8-hp preproduction models built with nova engine.
62	1937	1937	621,000	621,078	1st production 'Baby' type
L	1937	1938	621,079	622,563	Unstyled model.
L	1938	1941	625,000	634,840	Styled 3.25″ × 4″ engine.
	1941	1946	640,000	642,038	Styled 3.25″ × 4″ engine.
LI	1938 / 1942	1941 / 1946	625,000 / 50,001	634,840 / 52,019	Industrial model.
LA	1940	1946	1,001	13,475	Industrial model 3.5″ × 4″ engine.
M	1947	1952	10,001	50,580	4″ × 4″ vertical engine.
MT	1949	1952	10,001	35,845	Tricycle version of M. Single wheel and wide axle.
MC	1949	1952	10,001	16,309	Crawler—3 track rollers.
R diesel	1949	1954	1,000	22,293	Deere's first diesel. 5.75″ × 8″ main engine.
40S	1953	1956	60,001	77,906	Standard model. 4″ × 4″ vertical engine.
40U / 40U 2-row / 40 Special	1953 / 1955 / 1955	1956 / 1956 / 1956	60,001	63,140	Utility model. 4″ × 4″ engine. / 2-row extra wide & low / High clearance—26½″.
40 Hi-Crop	1954	1956	60,001	60,060	High clearance—32″.
40T	1953	1956	60,001	75,531	Tricycle row-crop, twin, single, or wide axle.
40C	1953	1956	60,001	77,906	4- or 5-roller crawler.
50	1952 / 1955	1956 / 1956	5,000,001	5,033,751	4¹¹⁄₁₆″ × 5½″ engine, tricycle type twin, single, wide axle. / LP-gas engine option added.
60	1952	1956	6,000,001	6,064,096	5½″ × 6¾″ engine, tricycle type, twin, single, wide axle; gas, LP-gas or all-fuel.
60S	1952	1954	6,000,001	6,042,732	Standard low-seat type
60S	1954	1956	6,043,000	6,064,096	Standard high-seat type
60-O	1952	1956	6,000,001	6,064,096	Orchard model.
60H	1952	1956	6,000,001	6,064,096	Hi-Crop, 32″ clearance.

JOHN DEERE TRACTOR MODEL PRODUCTION YEARS, (cont'd)

Model	Year Start	Finish	Serial Number First	Last	Features & Notes
70	1953	1956	7,000,001	7,043,757	5⅞″ × 7″ gas or LP-gas engine, 6⅛″ × 7″ all-fuel. Twin or single wheel or wide axle.
70D	1954	1956	7,000,001	7,043,757	6⅛″ × 6⅜″, diesel engine. Twin or single wheel or wide axle.
70H	1953	1956	7,000,001	7,043,757	Hi-Crop, 32″ clearance.
70S	1953	1956			Standard 4-wheel, all engine options.
80	1955	1956	8,000,001	8,003,500	6⅛″ × 8″ diesel engine, standard tread.
320	1956	1958	320,001	322,566	
320U	1956	1958	325,001	325,518	4″ × 4″ gas engine, standard one-row.
320S	1956	1958	325,001	325,517	
420	1956	1958	80,001	136,868	4¼″ × 4″ gas or all-fuel engine, standard 4-wheel.
420U	1956	1958	80,001	136,867	Utility low-built.
420RCU	1956	1958	80,001	136,864	Row-crop utility, 2-row.
420T	1956	1958	80,001	136,864	Tricycle, twin, single wheel or wide axle.
420H	1956	1958	80,001	136,856	Hi-Crop, 32″ clearance.
420S	1956	1958	80,001	136,866	High-built, 26″ clearance.
420C	1956	1958	80,001	136,795	4- or 5-roller crawler.
520	1956	1958	5,200,000	5,213,189	4¹¹⁄₁₆″ × 5½″ gas, LP-gas or all-fuel engine. Twin, single wheel or wide axle.
620	1956	1958	6,200,000	6,222,686	5½″ × 6⅜″ gas, LP-gas, or all-fuel engine. Twin or single wheel, fixed 38″ or adjustable wide front axle.
620S	1956	1958	6,200,195	6,208,866	Standard 4-wheel.
620-O	1957	1960	6,200,000	6,223,247	Grove/orchard model.
620H	1957	1958	6,201,868	6,208,055	Hi-Crop, 32″ clearance. 48″ between final drives.
720	1956	1958	7,200,000	7,229,002	7,229,002 was ex 7,215,262 6″ × 6⅜″ gas, LP-gas, or all-fuel engines. Twin, single, or wide front axle.
720D	1956	1958			6⅛″ × 6⅜″ diesel engine.
720S	1956	1958	7,200,000	7,229,002	Standard 4-wheel.
720H	1956	1958	7,200,649	7,228,603	Hi-Crop. As 620 dimensions.
820	1956	1958	8,200,000 / 8,200,003	8,200,001 / 8,207,080	Standard diesel 6⅛″ × 8″ engine.
330S	1958	1960	330,001	331,091	As "20" series. Modified styling.
330U	1958	1960	330,001	331,088	As "20" series. Modified styling.
430	1958	1960	140,001	160,994	As "20" series. Modified styling.
430U	1958	1960	140,001	161,075	As "20" series. Modified styling.
430RCU	1958	1960	140,001	161,016	As "20" series. Modified styling.
430T	1958	1960	140,001	161,096	As "20" series. Modified styling.
430H	1958	1960	140,001	160,996	As "20" series. Modified styling.
430Sp.	1958	1960	140,001	160,972	As "20" series. Modified styling.
430C	1958	1960	140,001	161,072	As "20" series. Modified styling.
435	1958	1960	435,001 / 439,588	439,450 / 439,626	G.M. 2-cyl. diesel. 3⅞″ × 4½″ engine.
530	1958	1960	5,300,000	5,309,814	As "20" series. Modified styling.
630	1958	1960	6,300,000	6,318,206	As "20" series. Modified styling.
630S	1958	1960			As "20" series. Modified styling.
630H	1958	1959			As "20" series. Modified styling.
730	1958	1960	7,300,000	7,330,358	As "20" series. Modified styling.
703S	1958	1960			As "20" series. Modified styling.
730H	1958	1960			As "20" series. Modified styling.
830	1958	1960	8,300,000	8,306,891	As "20" series. Modified styling.
840	1958	1960	8,400,000	8,400,848	Industrial offset model.

Model	Code	Beginning Serial No.	Date	Ending Serial No.	Date
D	101	30,401	11/30/23		
		30,402	11/30/23		
		30,403	7/11/23		
		30,404	6/16/23		
		30,405	6/ 6/23		
		30,406	6/16/23	30,410	6/16/23
		30,411	6/30/23		
		30,412	6/30/23		
		30,413	11/30/23		
		30,414	1/16/24		
		30,415	7/ 1/23	30,448	11/23/23
		30,449	9/ 8/23		
		30,450	1/25/24		
D	101	30,451	1/16/24	31,319	6/12/24
		31,413	11/ 1/24	53,387	9/11/27
D	103	53,388	9/12/27	109,943	11/10/30
D Crawler		107,001	7/ 9/30	107,048	7/22/30
D	108	109,944	11/11/30	119,071	11/21/34
D	116	119,100	11/27/34	130,651	11/30/36
				1 only 130,653	4/26/38
D	116	130,700	11/ 5/36	143,568	3/29/39
DI	124	125,581	12/30/35	149,187	8/28/40
D	147	143,800	4/ 8/39	191,670	7/ 3/53
C		101	3/11/27	125	5/ 5/27
		200,001	8/23/27	200,069	10/ 5/27
		200,080	12/22/27		
		200,099	10/24/27		
		200,102	12/30/27		
		200,106	1/12/28	200,110	1/19/28
GP	105	200,111	3/14/28	200,202	4/20/28
		200,211	8/ 4/28	223,801	8/15/30
GP Tricycle		200,264	8/18/28	204,213	4/15/29
		223,803	1/ 8/31	230,745	2/28/35
GPO	109	15,000	4/ 2/31	15,732	4/ 4/35
GPWT	106	400,000	11/12/28	404,767	2/ 9/32
	110	404,810	2/10/32	405,254	11/ 1/33
P	107	5,000	1/30/30	5,202	8/19/30
AA	112	410,000	4/ 8/33	410,007	6/22/33
A		410,008	4/ 8/34	414,808	2/11/35
		414,809	2/11/35	476,217	6/13/38
AN	123	420,660	8/ 5/35	475,945	6/ 1/38
AW	118	417,450	5/10/35	475,946	6/ 2/38
ANH	131	469,668	12/16/37	476,221	6/13/38
AWH	132	469,807	12/21/37	476,218	6/15/38
AR	117	250,000	4/25/35	259,334	10/23/40
AO	121	250,075	5/22/35	253,482	10/ 5/36
AI	125	252,334	4/27/36	259,335 & 260,329	10/30/40—3/20/41
B	114	1,000	10/ 2/34	42,133	6/24/37
		42,200	6/24/37	58,246	6/14/38
BN	115	4,244	4/ 1/35	58,197	6/13/38
BW	119	4,431	4/11/35	57,759	5/31/38
BNH	129	46,175	10/ 1/37	58,176	6/ 8/38
BWH	130	51,679	12/15/37	58,095	6/ 8/38
BR	120	325,000	9/24/35	328,884	6/ 8/38
BO	122	325,084	9/27/35	328,890	6/24/38
BI	126	326,016	4/ 1/36	328,889	6/14/38
BR	138	329,000	6/14/38	337,514	1/28/47
BO	139	329,082	6/16/38	337,506	1/15/47
BI	140	329,083	6/16/38	332,157	2/27/41
Y		Not numbered	1936	26 made	1936
62		62.1000	1/37	62.1078	1937
L		62.1079	8/37	62.2563	1938
STYLED:					
L		625,000	8/15/38	634,840	8/25/41
L (Deere engine)		640,000	7/17/41	642,038	7/ 9/46
LA		1,001	8/ 2/40	13,475	8/12/46
LI		50,001	3/18/41	52,019	4/ 5/46
UNSTYLED:					
G	127	1,000	5/17/37	4,250	1/29/38
		4,251	1/19/38	7,000	7/16/38
		7,100	7/18/38	12,192	12/22/41
STYLED:					
GM	166	13,000	2/18/42	22,112	3/10/47
G		23,000	3/ 7/47	64,530	2/19/53
GN	175	24,382	5/15/47	64,358	2/ 9/53
GW	176	24,377	5/16/47	64,526	2/18/53
GH	184	46,894	8/ 3/50	64,163	1/14/53
A	141	477,000	8/ 1/38	487,247	10/10/39
		487,248	3/24/41	487,249	3/24/41
AN	142	477,174	8/ 4/38	487,130	8/21/39
AW	144	477,231	8/ 8/38	487,234	10/10/39
ANH	143	477,241	8/ 9/38	486,906	6/15/39
AWH	145	477,230	8/ 8/38	485,843	4/21/39
A	148	488,000	9/18/39	498,535	4/22/41
AN	149	488,272	10/12/39	498,529	9/ 5/40
AW	151	488,277	10/11/39	498,532	9/ 7/40
ANH	150	488,209	10/ 4/39	497,668	7/17/40
AWH	152	488,284	10/12/39	498,164	8/13/40
A	160	499,000	9/12/40	583,326	2/ 4/47
AN	161	499,139	9/11/40	583,301	2/ 3/47
AW	163	499,147	9/10/40	583,253	1/31/47
ANH	162	499,145	9/12/40	583,194	1/31/47
AWH	164	499,169	9/11/40	582,462	1/15/47
A	169	584,000 From 648,000	3/31/47	180 703,384	5/12/52
AN	170	585,427 From 648,000	5/15/47	181 703,307	5/ 8/52
AW	171	585,426 From 648,000	5/15/47	182 703,306	5/ 8/52
AH	183	665,665	7/10/50	702,428	4/25/52

TRACTOR MODEL BUILD DATES FOR FIRST AND LAST MADE (Cont'd).

Model	Code	Beginning Serial No.	Date	Ending Serial No.	Date
AOS	128	1,000	11/23/86	1,497	10/ 4/37
		1,525	10/ 6/37	1,693	5/25/38
		1,697	12/21/37	1,771	10/18/39
		1,800	10/31/39	1,836	3/28/40
		1,850	4/30/40	1,891	10/28/40
AR	159	260,057	12/10/40	270,678	11/ 9/48
AO	165	260,000	11/27/40	270,679	11/ 9/48
AR	178	272,000	6/ 7/49	284,074	5/ 6/53
AO	179	272,112	7/ 7/49	284,073	5/ 6/53
H	146	1,000	10/29/38	61,116	2/ 6/47
HN	153	14,874	2/19/40	61,113	2/ 6/47
HNH	167	30,172	3/11/41	42,726	1/23/42
HWH	168	29,982	3/ 6/41	42,695	1/22/42
B	133	60,000	6/16/38	95,184	8/29/40
BN	134	60,097	7/26/38	95,200	10/11/40
BW	136	60,563	8/ 9/38	95,201	10/17/40
BNH	135	60,335	8/ 2/38	94,552	7/17/40
BWH	137	60,531	8/10/38	94,742	7/25/40
B	154	96,000	9/ 4/40	200,247	5/12/47
BN	155	96,210	9/11/40	200,223	12/28/46
BW	157	96,209	9/11/40	200,209	12/30/46
BNH	156	96,198	9/11/40	199,802	12/19/46
BWH	158	96,208	9/11/40	199,299	12/ 3/46
B	172	201,000	2/ 4/47	310,772	6/ 2/52
BN	173	203,247	3/21/47	310,747	5/29/52
BW	174	203,736	3/31/47	310,748	6/ 2/52
M		10,001	1947	55,799	9/ 8/52
MT		10,001	12/21/48	40,472	9/ 4/52
MC		10,001	12/28/48	20,509	9/ 4/52
MI		10,001	1/ 2/49	65,208	10/17/55
R	177	1,000	2/ 1/49	22,293	11/ 1/54
NUMBERED SERIES:					
40S		60,001	1/ 9/53	71,814	10/18/55
40U		60,001	11/10/52	71,689	10/19/55
40T		60,001	10/30/52	77,906	11/21/55
40V Special		60,001	11/ 4/54	60,329	10/14/55
40C		60,001	2/12/53	71,689	10/19/55
40W R-C Utility		60,001	2/10/55	61,758	10/21/55
50		5,000,001	7/22/52	5,033,751	5/14/56
60		6,000,001	3/12/52	6,063,853	7/20/56
60S Low seat		6,000,001	3/12/52	6,042,732	9/12/54
60S High seat		6,043,000	1/12/54	6,064,096	5/ 1/57
60-O		6,054,723	8/19/55	6,064,092	2/ 5/57
				& 6,064,096	5/ 1/57
70	1401	7,000,001	3/27/53	7,043,755	6/22/56
70S	1350	7,003,593	9/10/53	7,043,757	7/20/56
70H	1400	7,005,300	10/21/53	7,043,636	6/21/56
70D	diesel	7,017,500	10/27/54	7,043,757	7/20/56
80		8,000,001	6/27/55	8,003,500	7/11/56
320S		320,001	6/25/56	325,517	7/30/58
320U		320,017	8/ 1/56	325,518	8/ 2/56
420S		80,032	11/ 7/55	136,866	8/ 4/58
420U		80,027	11/14/55	136,867	8/ 2/58
420T		80,001	11/ 2/55	136,864	7/29/58
420W		80,179	11/17/55	136,868	7/31/58
420V		80,091	11/ 7/55	135,142	6/19/58
420C		80,002	11/ 2/55	136,795	7/29/58
520		5,200,000	5/ 4/56	5,213,189	8/ 1/58
620		6,200,000	6/ 7/56	6,222,686	7/25/58
620S		6,200,195	7/20/56	6,208,866	2/28/57
620H		6,201,868	9/20/56	6,208,055	2/11/57
620-O		6,200,000	1957	6,223,247	2/ 1/60
720		7,200,000	7/12/56	7,229,001	7/28/58
		(7,215,262	9/ 4/57) became	7,229,002	7/30/58
720S		7,200,125	8/21/56	7,228,920	7/29/58
720H		7,200,649	8/30/56	7,228,603	7/23/58
820		8,200,000	7/10/56	8,207,078	7/23/58
330S		330,001	8/ 2/58	331,091	3/22/60
330U		330,008	8/ 1/58	331,088	2/25/60
430S		140,027	8/ 2/58	160,994	2/24/60
430U		140,003	7/31/58	161,075	3/ 4/60
430T		140,006	7/28/58	161,096	3/ 3/60
430W		140,013	8/ 1/58	160,984	2/26/60
430V		140,557	9/ 4/58	160,582	1/29/60
430C		140,001	8/ 1/58	161,072	3/ 4/60
435-Diesel		435,001	1/31/59	439,626	4/13/60
530		5,300,000	8/ 4/58	5,309,814	9/27/60
530-LP		5,300,022	8/ 4/58	5,309,438	3/ 1/60
530-Allfuel		5,300,581	9/12/58	5,309,666	3/21/60
630		6,300,000	8/ 5/58	6,318,206	4/20/60
630S		6,300,088	8/ 5/58	6,317,201	2/ 5/60
630H		6,300,687	9/ 9/58	6,315,983	12/28/59
730		7,300,000	8/ 4/58	7,330,358	3/ 1/60
730S		7,300,028	8/ 4/58	7,330,312	3/ 1/60
730H		7,300,080	8/ 5/58	7,326,068	2/11/60
830		8,300,000	8/ 4/58	8,306,891	7/14/60
840		8,400,000	10/16/58	8,400,848	9/27/60

A-6   MODEL "C" TRACTORS— ORIGINAL SERIAL NOS.

Original Serial Nos.	Date Shipped	Type
1	4/21/27	Tricycle
2	2/ 2/27	4-wheel
3	12/24/28	4-wheel
4	6/19/28	4-wheel
9	4/ 2/28	Tricycle
101	3/11/27	4-wheel
102	6/24/27	4-wheel
103	3/24/27	4-wheel
104	4/15/27	4-wheel
105	3/31/27	4-wheel
106	4/22/27	4-wheel
107	4/ 1/27	4-wheel
108	4/11/27	4-wheel
109	4/11/27	4-wheel
110	4/15/27	4-wheel
111	4/15/27	4-wheel
112	4/22/27	4-wheel
113	4/15/27	4-wheel
114	4/22/27	4-wheel
115	4/18/27	4-wheel
116	4/22/27	4-wheel
117	4/22/27	4-wheel
118	4/22/27	4-wheel
119	4/29/27	4-wheel
120	4/29/27	4-wheel
121	5/ 1/27	4-wheel
122	5/10/27	4-wheel
123	?	4-wheel
124	5/ 5/27	4-wheel
125	5/ 5/27 (shipped)	4-wheel
200,001-200,069	8/23—10/ 5/27	4-wheel
200,080	12/22/27	4-wheel
200,099	10/24/27	4-wheel
200,102	12/12/27	4-wheel
200,103	10/10/28	4-wheel
200,106	1/12/28	4-wheel
200,107	1/12/28	4-wheel
200,108	1/12/28	4-wheel
200,109	1/19/28	4-wheel
200,110	1/19/28	4-wheel

A-7   MODEL "GP" TRACTORS— NEW SERIAL NOS.

New Number	Original C Number	Build Date	Type
200,111	200,015	3/15/28	(4-wheel to
200,112	New	10/19/28	200,263—then
200,113	New	3/17/28	see Tricycle
200,114	200,043	3/22/28	Tractor List,
200,115	New	3/24/28	A-8.)
200,116	200,050	3/22/28	
200,117	200,033	3/26/28	
200,118	122	3/21/28	
200,119	200,056	3/22/28	
200,120	200,054	3/26/28	
200,121	104	3/26/28	
200,122	200,002	3/28/28	
200,123	109	3/26/28	
200,124	200,006	3/26/28	
200,125	200,057	3/26/28	
200,126	200,036	3/29/28	
200,127	106	3/29/28	
200,128	200,014	3/29/28	
200,129	200,032	3/29/28	
200,130	113	3/29/28	

A-7 (cont'd)   MODEL "GP" TRACTORS— NEW SERIAL NOS.

New Number	Original C Number	Build Date	Type
200,131	108	3/29/28	
200,132	116	3/29/28	
200,133	200,028	3/31/28	
200,134	200,013	3/31/28	
200,135	200,030	3/31/28	
200,136	120	4/ 3/28	
200,137	200,035	3/31/28	
200,138	200,051	3/31/28	
200,139	200,020	3/31/28	
200,140	200,001	3/31/28	
200,141	200,053	3/31/28	
200,142	112	4/ 2/28	
200,143	200,059	3/31/28	
200,144	200,017	4/ 3/28	
200,145	124	4/ 3/28	
200,146	200,026	4/ 3/28	
200,147	200,023	4/ 3/28	
200,148	200,025	4/ 3/28	
200,149	107	4/ 4/28	
200,150	200,037	4/ 3/28	
200,151	121	4/ 8/28	
200,152	110	4/ 5/28	
200,153	115	4/ 5/28	
200,154	New	4/ 5/28	
200,155	200,029	4/ 5/28	
200,156	200,019	4/ 5/28	
200,157	200,007	4/ 5/28	
200,158	111	4/ 6/28	
200,159	200,060	4/ 6/28	
200,160	200,063	4/ 6/28	
200,161	New	4/ 7/28	
200,162	New	4/ 7/28	
200,163	200,040	4/ 7/28	
200,164	New	4/ 7/28	
200,165	New	4/ 9/28	
200,166	200,066	4/ 9/28	
200,167	New	4/ 9/28	
200,168	200,069	4/10/28	
200,169	New	4/10/28	
200,170	200,068	4/10/28	
200,171	118	4/11/28	
200,172-174	New	4/12/28	
200,175-177	New	4/13/28	
200,178	200,034	4/13/28	
200,179	New	4/14/28	
200,180		4/17/28	
200,181		4/14/28	
200,182	200,008	4/13/28	
200,183	New	4/14/28	
200,184		4/17/28	
200,185		4/18/28	
200,186		4/18/28	
200,187-191	Service Note	4/18/28	
200,192-196	Service Note	4/20/28	
200,197-199	Service Note	4/21/28	
200,200	Service Note	4/23/28	
200,201	New	4/23/28	
200,202	Service Note	4/25/28	
200,203—200,210	Nos. not used.		
200,211-230,745	New series. Except GP Tricycle Tractors (see A-8)	8/ 4/28—2/28/35	

383

A-8 "GP TRICYCLE" TRACTORS

MODEL	Build Date
200,264	8/18/28
200,272	8/14/28
200,287	8/17/28
200,297	8/17/28
200,346	8/22/28
200,366	9/ 1/28
200,377	9/ 5/28
200,400	8/29/28
200,403	8/29/28
200,406	8/29/28
200,437	8/31/28
200,438	9/ 1/28
200,458	9/ 5/28
200,488	9/ 6/28
202,380	1/19/29
202,401	1/31/29
203,784	4/26/29
203,970	4/27/29
203,989	4/11/29
204,072	3/ 6/29
204,077	3/22/29
204,172	4/19/29
204,213	4/15/29

A-9 MODEL "AA"— "A" SERIAL NUMBERS

MODEL	"AA" Numbers	New "A" Numbers	Date Originally Shipped
AA-1	410,000	412,036	4/ 4/33
*AA-3	410,001	Scrapped	4/ 8/33
AA-1	410,002	412,005	4/29/33
AA-1	410,003	411,928	4/11/33
AA-1	410,004	412,134	5/ 3/33
*AA-3	410,005	412,102	4/ 8/33
AA-1	410,006	412,562	6/10/33
AA-1	410,007	410,007 or 412,760	6/22/33

			Date rebuilt & shipped
Pre-A	410,008	415,178	3/30/35
Pre-A	410,009	415,197	3/30/35
Pre-A	410,010	410,010	3/25/36
Pre-A	410,011	412,866	4/ 8/36
A	410,012	First A shipped 3/17/34	
A	410,022	First "A" built on rubber tires. 4/9/34	

*Model AA-3 cancelled 7/31/33.

Model	Years	SP/PT-GD/EF/PTO	Cyl. Width	Header Size
Preproduction	1925-1926	PT, GD	24"	12', 16'
1	1928-1929	PT, EF	24"	8', 10'
2	1927-1929	PT, EF	24"	12', 16'
3	1928-1932	PT, EF	30"	12'
4 (not produced)	1929	PT, EF	24"	12'
5	1929-1934	PT, EF	24"	10', 12'
5A	1934-1941	PT, EF	24"	10', 12'
6	1936-1939	PT, PTO, EF	24"	6'
7	1932-1940	PT, EF	24"	8'
7A	1941-1942	PT, EF	24"	8'
9	1939-1942 1945-1946	PT, PTO, EF	30"	12'
10	1939	PT, PTO	40"	3'6"
11	1939	PT, PTO	50"	5'
12	1939	PT, PTO	60"	6'
10A	1940	PT, PTO	40"	3'6"
10AW Experimental	1944	PT, PTO	40"	5'
11A	1940-1942	PT, PTO	50"	5'
12A	1940-1945 1946-1952	PT, PTO PT, PTO, EF	60"	6'
17	1932-1942 1945-1948	PT, EF	30"	12', 16'
25	1952-1955	PT, PTO, EF	60"	6', 7'
30	1956-1960	PT, PTO, EF	60"	7'
33 Hillside	1940-1942	PT, EF	20"	10'
35 Hillside	1937-1942	PT, EF	24"	12', 14'
36 Level Land	1927-1942	PT, EF	30"	12', 16½', 20'
36A Extreme Hillside	1927-1942	PT, EF	30"	16½'
36B Medium Hillside	1936-1942	PT, EF	30"	12', 16½', 20'
36 Level Land	1943-1951	PT, EF	30"	16½', 20'
36 Hillside	1943-1951	PT, EF	30"	12', 16½', 20'
40 Hi-Lo	1960-1966	SP	24⅔"	8', 10'
42 Hi-Lo	1961-1966	PT, PTO	24⅔"	9'
45	1954-1959	SP	26"	8', 10'
45 Hi-Lo	1960-1969	SP	26"	8', 10'
55 Experimental	1944	SP	30"	12'
55C Experimental	1945	SP	30"	12'
55X Experimental	1945	SP	30"	12'
55	1946-1959	SP	30"	12', 14'
55R & RC Rice	1946-1959	SP	30"	12', 14'
55H Hillside	1954-1960	SP	30"	14'
55 Hi-Lo	1960-1969	SP	30"	12', 13', 14',
55R & RC Rice Hi-Lo	1960-1969	SP	30"	& 15'
65	1949-1959	PT, PTO, EF	30"	12'
95 & 95R & RC	1958-1959	SP	40"	14', 16', 18'
95H Hillside	1958-1959	SP	40"	16', 18'
95, 95R, & RC Hi-Lo	1960-1969	SP	40"	12' to 20'
95H Hi-Lo Hillside	1960-1969	SP	40"	16', 18'

Modified Table from The Grain Harvesters
Graeme R. Quick, author, published by the American Society of Agricultural Engineers, 2950 Niles Rd., St. Joseph, Michigan 49085-9659 USA.

Tractor Specifications

JOHN DEERE TRACTOR SPECIFICATIONS

MODEL	DATE	CAPACITY IN PLOWS Drawbar H.P.	TRANSMISSION SPEEDS miles per hour						REVERSE	CYLINDERS	BORE	ENGINE STROKE (inches)	RPM
			1	2	3	4	5	6					
WB/Overtime R	6-17	3 (14")	2½						2½	2	6	7	750
WB/Overtime N	1920	3 (14")	2¼	3					2¼	2	6½	7	750
AWD	1919	3 (14")	2	2⅝					2 & 2⅝	4	4½	6	
D	'23-'27	3 (14")	2½	3¼					2	2	6½	7	800
D	'28-'30	3 (14")	2½	3¼					2	2	6¾	7	800
D	'31-'34	3 (14")	2½	3¼					2	2	6¾	7	900
D	'35-'53	3 (14")	2¼	3¼	4				1½	2	6¾	7	900
C/GP	'28-'30	2 (14")	2¼	3	4				1¾	2	5¾	6	950
GP	'31-'35	2 (14")	2¼	3	4				1¾	2	6	6	950
GPWT	'29-'30	2 (14")	2¼	3	4⅛				1¾	2	5¾	6	950
GPWT	'31-'32	2 (14")	2¼	3	4⅛				1¾	2	6	6	950
GPWT O/H	'32-'33	2 (14")	2¼	3	4⅛				1¾	2	6	6	950
GPO	'30-'35	2 (14")	2¼	3	4				1¾	2	6	6	950
A	12-33	2 (14")	2⅓	3	4¾	6¼			3½	2	5½	6½	975
B	2-36	2 (14")	2⅓	3	4¾	6¼			3½	2	4¼	5¼	1150
AR	12-40	2 (16")	2	3	4	6¼			3	2	5½	6¾	975
AO	12-40	2 (16")	2	3	4	6¼			3	2	5½	6¾	975
AI	5-38	2 (16")	2	3	4	6¼			3	2	5½	6½	975
BR	12-40	2 (14")	2	3¼	4¼	6¾			3½	2	4½	5½	1150
BO	12-40	2 (14")	2	3¼	4¼	6¾			3½	2	4½	5½	1150
BI	5-38	2 (14")	2	3¼	6¼	9¾			3½	2	4¼	5¼	1150
BO-Lindeman	1939	3 (14")	1¼	2	2¾	4			2	2	4½	5½	1150
A	6-38	2 (16")	2⅓	3	4	5¼			3¾	2	5½	6½	975
B	6-38	2 (14")	2⅓	3	4	5¼			3¾	2	4½	5½	1150
G	6-38	3 (14")	2¼	3¼	4¼	6			3	2	6⅛	7	975
H	3-39	1 (16")	2½	3½	5¾				1¾	2	3⁹⁄₁₆	5	1400
62/L/LI	9-41	1 (12")	2	3½	6½				2½	2	3¼	4	1550
LA	9-41	1 (16")	2½	3¾	8				2½	2	3½	4	1850
GM	4-41	3 (14")	2½	3½	4½	6⅓	8½	12	3¼	2	6⅛	7	975

MODEL	DATE	DRAWBAR HORSEPOWER	1	2	3	4	5	6	REVERSE	CYLINDERS	BORE	ENGINE STROKE (inches)	RPM
A	4-51	30.98	1½	2½	3½	4½	6	11	3	2	5½	6¾	975
A	4-51	35.30	1½	2½	3½	4½	6	11	3	2	5½	6¾	975
B	4-51	22.19	1½	2½	3½	4½	5¾	10	2½	2	4¹¹⁄₁₆	5½	1250
B	4-51	25.50	1½	2½	3½	4½	5¾	10	2½	2	4¹¹⁄₁₆	5½	1250
G	4-51	36.01	2½	3½	4½	6½	8¾	12½	3½	2	6⅛	7	975
M	1-49	18.15	1⅝	3⅛	4¼	10			1⅝	2	4	4	1650
MT	11-50	18.8	1⅝	3⅛	4¼	10			1⅝	2	4	4	1650
MC	7-50	18.3	.8	2.2	2.9	4.7			1	2	4	4	1650
R	7-51	45.69	2⅛	3⅓	4¼	5⅓	11½		2½	2	5¾	8	1000
AR styled	12-51	27.13	1⅓	2½	3¼	4½	6¼	11	2¾	2	5½	6¾	975
AR styled	12-51	34.88	1⅓	2½	3¼	4½	6¼	11	2¾	2	5½	6¾	975
AO styled	12-51	27.13	1⅓	2½	3¼	4½	6¼	11	2¾	2	5½	6¾	975
AO styled	12-51	34.88	1⅓	2½	3¼	4½	6¼	11	2¾	2	5½	6¾	975
AH	5-51	26.83	1½	2½	3½	4½	6½	11¼	3	2	5½	6¾	975
AH	5-51	35.30	1½	2½	3½	4½	6½	11¼	3	2	5½	6¾	975
GH	5-51	36.01	2¼	3¼	4¼	6	8¼	11¾	3	2	6⅛	7	975
40T	3-54	22.9	1⅝	3⅛	4¼	12			2½	2	4	4	1850
40S	3-54	22.4	1⅝	3⅛	4¼	12			2½	2	4	4	1850
40C	9-53	20.1	.82	2.21	2.95	5.31			1.64	2	4	4	1850
50	6-53	27.49	1½	2½	3½	4½	5¾	10	2½	2	4¹¹⁄₁₆	5½	1250

MODEL	FUEL	FUEL TANK CAPACITY		COOLING CAPACITY	STANDARD WHEEL SIZE (inches)		DIMENSIONS (inches)			WEIGHT (lbs.)
		MAIN	AUXILIARY		FRONT	REAR	LENGTH	WIDTH	HEIGHT	
		(US gallons)		(US gallons)						
WB/Overtime R	K	20	1	8½	28×6	52×10	143	72	63*	
WB/Overtime N	K	20	1	8½	28×6	52×12	132	72	63*	6,183
AWD	K				36×8	40×20	150	76	57	4,600
D	K	18	2½	13	28×5	46×12	109	63	56	4,000
D	K	18	2½	13	28×5	46×12	109	63	56	4,164
D	K	25	1½	10	28×6	46×12	130	66½	58¼	5,114
D	K	25	1½	10	28×6	46×12	130	66½	61¼	5,269
C/GP	K	16	2	9	24×6	42¾×10	112	60	55½	3,600
GP	K	16	2	9	24×6	42¾×10	112	60	55½	3,600
GPWT	K	16	2	9	24×4	44×10	117½	85½	58	
GPWT	K	16	2	9	24×4	44×10	117½	85½	58	
GPWT O/H	K	14	1	9	24×4	44×10	129⅛	85¾	61½	
GPO	K	16	2	9	24×6	42¾×10	121	64	49	4,250
A	K	14	1	9½	24×4	50×6	124	86	60	3,525
B	K	12	1	5½	22×3¼	48×5¼	120½	85	56	2,763
AR	K	16	1	8	6.00×16	11.25×24	124	64⅛	55	4,384
AO	K	16	1	8	6.00×16	11.25×24	124	64⅛	55	4,474
AI	K	16	1	8	6.00×16	11.25×24	119½	64½	54½	4,680
BR	K	12	1	6	5.50×16	9.00×28	117¾	52¾	50½	3,375
BO	K	12	1	6	5.50×16	9.00×28	117¾	52½	50½	3,441
BI	K	10	1	5½	5.50×16	9.00×28	115	53¾	51¾	3,620
BO-Lindeman	K	13½	1	6	tracks—10″ shoes		86	53	50½	4,420
A	K	14	1	8	24×4	50×6	133	83	62½	3,783
B	K	13½	1	6	22×3¼	48×5¼	125½	83½	57	2,878
G	K	17	1½	11	24×5	51½×7	135	84	61½	4,400
H	K	7½	⅞	5½	4.00×15	8×32	111¼	79	52	2,063
62/L/LI	G	6		2½	4.00×15	6.00×22	91	49	57	1,515
LA	G	8		2½	5.00×15	8.00×24	93	47	60	2,200
GM	K	17	1½	13	6.00×16	11-38	137⁷⁄₁₆	84¾	65⅞	5,100
A	K	14	1	8¾	5.50×16	11-38	134	86⅜	63⅞	4,909
A	G	18		8¾	5.50×16	11-38	134	86⅜	63⅞	4,909
B	K	12	2	7	5.50×16	10-38	132¼	86¹¹⁄₁₆	59⅝	4,052
B	G	12		7	5.50×16	10-38	132¼	86¹¹⁄₁₆	59⅝	4,052
G	K	17	1½	13	6.00×16	12-38	137⁷⁄₁₆	84¾	65⅞	5,624
M	G	10		3½	4.00×15	8-24	110	51	56	2,560
MT	G	9		3½	5.00×15	9-34	125⅜	89¾	58¾	2,900
MC	G	9		3½	tracks—10″ shoes		102	67	50½	4,000
R	D	22	¼	13⅝	7.50×18	14-34	147	79½	78⅛*	7,400
AR styled	K	20	1	8¾	6.00×16	13-26	125½	71½	57	5,245
AR styled	G	20		8¾	6.00×16	13-26	125½	71½	57	5,245
AO styled	K	20	1	8¾	6.00×16	13-26	125½	75⅝	57	5,429
AO styled	G	20		8¾	6.00×16	13-26	125½	75⅝	57	5,429
AH	K	14	1	8¾	7.50×20	11-38	152½	83	78	6,468
AH	G	18		8¾	7.50×20	11-38	152½	83	78	6,468
GH	K	16	1½		7.50×20	12-38	154	83	78¾	6,940
40T	G	10½		3½	5.00×15	9-34	130⅝	89¾	58¾	3,000
40S	G	10½		3½	5.00×15	9-24	114¾	55½	56	2,750
40C	G	11		3½	tracks—12″ shoes		102	67	50½	4,000
50	K	15½	2	7	5.50×16	10-38	132¾	86⅝	59⅞	4,435
50	G	15½		7	5.50×16	10-38	132¾	86⅝	59⅞	4,435

JOHN DEERE TRACTOR SPECIFICATIONS

MODEL	DATE	CAPACITY IN PLOWS Drawbar H.P.	1	2	3	4	5	6	REVERSE	CYLINDERS	BORE	ENGINE STROKE (inches)	RPM
60	6-53	30.1	1½	2½	3½	4½	6¼	11	3	2	5½	6¾	975
60	6-53	36.9	1½	2½	3½	4½	6¼	11	3	2	5½	6¾	975
60	6-53	38.1	1½	2½	3½	4½	6¼	11	3	2	5½	6¾	975
70	6-53	41.0	2½	3½	4½	6½	8¾	12½	3¼	2	5⅞	7	975
70	6-53	44.2	2½	3½	4½	6½	8¾	12½	3¼	2	5⅞	7	975
70	6-53	46.1	2½	3½	4½	6½	8¾	12½	3¼	2	5⅞	7	975
70D	12-54	45.66	2½	3⅔	4¾	6⅔	9	12¾	3⅓	2	6⅛	6⅜	1125
70D standard	12-54	45.66	2½	3½	4½	6¼	8½	12	3¼	2	6⅛	6⅜	1125
80	12-55	61.76	2⅓	3½	4½	5⅓	6¾	12¼	2⅔	2	6⅛	8	1125
320/330S	12-55	22.4	1⅝	3⅛	4¼	12			1⅝	2	4	4	1850
320/330U	12-55	22.4	1⅝	3⅛	4¼	12			1⅝	2	4	4	1850
420/430S	12-55	21.89	1⅝	3⅛	4¼	12	5th		2½	2	4¼	4	1850
420/430S	12-55	27.08	1⅝	3⅛	4¼	12	gear		2½	2	4¼	4	1850
430S	9-58	⅔ plow	1⅝	3⅛	4¼	12	optional		2½	2	4¼	4	1850
420/430T	12-55	21.89	1⅝	3⅛	4¼	12	5th		2½	2	4¼	4	1850
420/430T	12-55	27.08	1⅝	3⅛	4¼	12	gear		2½	2	4¼	4	1850
430T	10-58	⅔ plow	1⅝	3⅛	4¼	12	optional		2½	2	4¼	4	1850
420/430RCU	5-56	21.89	1⅝	3⅛	4¼	12	5th		2½	2	4¼	4	1850
420/430RCU	5-56	27.08	1⅝	3⅛	4¼	12	gear		2½	2	4¼	4	1850
430RCU	10-58	⅔ plow	1⅝	3⅛	4¼	12	optional		2½	2	4¼	4	1850
420/430C	12-55	24.12	1⅛	2¼	3	7¼	5th		1¾	2	4¼	4	1850
420/430C	12-55	¾ plow	1⅛	2¼	3	7¼	gear		1¾	2	4¼	4	1850
430C	9-58	¾ plow	1⅛	2¼	3	7¼	optional		1¾	2	4¼	4	1850
430HC	5-58	21.89	1⅝	3⅛	4¼	12	5th		2½	2	4¼	4	1850
430HC	5-58	27.08	1⅝	3⅛	4¼	12	gear		2½	2	4¼	4	1850
430HC	5-58	⅔ plow	1⅝	3⅛	4¼	12	optional		2½	2	4¼	4	1850
520/530	11-56	24.77	1½	2½	3½	4½	5¾	10	2½	2	4¹¹⁄₁₆	5½	1325
520/530	11-56	34.31	1½	2½	3½	4½	5¾	10	2½	2	4¹¹⁄₁₆	5½	1325
520/530	11-56	34.17	1½	2½	3½	4½	5¾	10	2½	2	4¹¹⁄₁₆	5½	1325
620/630RC	11-56	32.66	1½	2⅔	3⅔	4½	6½	11½	3	2	5½	6⅜	1125
620/630RC	11-56	44.16	1½	2⅔	3⅔	4½	6½	11½	3	2	5½	6⅜	1125
620/630RC	11-56	45.78	1½	2⅔	3⅔	4½	6½	11½	3	2	5½	6⅜	1125
620/630HC	5-58	32.66	1½	2⅔	3⅔	4⅔	6⅔	11⅔	3	2	5½	6⅜	1125
620/630HC	5-58	44.16	1½	2⅔	3⅔	4⅔	6⅔	11⅔	3	2	5½	6⅜	1125
620/630HC	5-58	45.78	1½	2⅔	3⅔	4⅔	6⅔	11⅔	3	2	5½	6⅜	1125
620/630S	7-58	32.66	1⅓	2½	3⅓	4¼	6	10½	2¾	2	5½	6⅜	1125
620/630S	7-58	44.16	1⅓	2½	3⅓	4¼	6	10½	2¾	2	5½	6⅜	1125
620/630S	7-58	45.78	1⅓	2½	3⅓	4¼	6	10½	2¾	2	5½	6⅜	1125
620 grove	2-57	4 plow	1½	2½	3½	4¼	6¼	10¾	2¾	2	5½	6⅜	1125
620 grove	2-57	4 plow	1½	2½	3½	4¼	6¼	10¾	2¾	2	5½	6⅜	1125
620 grove	2-57	4 plow	1½	2½	3½	4¼	6¼	10¾	2¾	2	5½	6⅜	1125
720/730RC	11-56	41.29	1⅓	2¼	3½	4⅓	5¾	11¼	3⅓	2	6	6⅜	1125
720/730RC	11-56	53.05	1⅓	2¼	3½	4⅓	5¾	11¼	3⅓	2	6	6⅜	1125
720/730RC	11-56	54.17	1⅓	2¼	3½	4⅓	5¾	11¼	3⅓	2	6	6⅜	1125
720D/730D RC	5-58	53.66	1⅓	2¼	3½	4⅓	5¾	11¼	3⅓	2	6⅛	6⅜	1125
720/730HC	5-58	41.29	1⅓	2¼	3⅓	4¼	5⅓	10½	3¼	2	6	6⅜	1125
720/730HC	5-58	53.05	1⅓	2¼	3⅓	4¼	5⅓	10½	3¼	2	6	6⅜	1125
720/730HC	5-58	54.17	1⅓	2¼	3⅓	4¼	5⅓	10½	3¼	2	6	6⅜	1125
720D/730D HC	5-58	53.66	1⅓	2¼	3⅓	4¼	5⅓	10½	3¼	2	6⅛	6⅜	1125
720/730S	7-58	41.29	1⅓	2¼	3½	4⅓	5½	11	3¼	2	6	6⅜	1125
720/730S	7-58	53.05	1⅓	2¼	3½	4⅓	5½	11	3¼	2	6	6⅜	1125
720D/730S	7-58	54.17	1⅓	2¼	3½	4⅓	5½	11	3¼	2	6	6⅜	1125
720D/730S	7-58	53.66	1⅓	2¼	3½	4⅓	5½	11	3¼	2	6⅛	6⅜	1125
820/830	10-57	69.66	2⅓	3½	4½	5⅓	6¾	12¼	2⅔	2	6⅛	8	1125
435	6-59	28.41	1⅞	3½	4¾	13½			2⅞	2	3.87	4.5	1850

MODEL	FUEL	FUEL TANK CAPACITY (US gallons) MAIN	AUXILIARY	COOLING CAPACITY (US gallons)	STANDARD WHEEL SIZE (inches) FRONT	REAR	DIMENSIONS (inches) LENGTH	WIDTH	HEIGHT	WEIGHT (lbs.)
60	K	20½	1	8¼	6.00 × 16	11-38	139	86⅝	65 9/16	5,300
60	G	20½		8¼	6.00 × 16	11-38	139	86⅝	65 9/16	5,300
60	LP	33		8¼	6.00 × 16	11-38	139	86⅝	65 9/16	5,600
70	K	24½	1¾	8½	6.00 × 16	12-38	136¼	86⅝	65 9/16	6,035
70	G	24½		8½	6.00 × 16	12-38	136¼	86⅝	65 9/16	6,035
70	LP	33		8½	6.00 × 16	12-38	136¼	86⅝	65 9/16	6,335
70D	D	20	¼	7	6.00 × 16	13-38	137	86⅝	88*	6,510
70D standard	D	20	¼	7	6.50 × 18	14-30	129⅞	86⅝	87¾*	7,205
80	D	32½	¼	8¾	7.50 × 18	15-34	142¾	79½	81*	7,850
320/330S	G	10½		3½	5.00 × 15	9-24	115¾	53½	55½	2,750
320/330U	G	10½		3½	5.00 × 15	9-24	119¼	55½	50¼	2,750
420/430S	K	10½	9/10	2¾	5.00 × 15	9-24	114¾	55½	55½	2,750
420/430S	G	10½		2¾	5.00 × 15	9-24	114¾	55½	55½	2,750
430S	LP	20		2¾	5.00 × 15	9-24	114¾	55½	55½	
420/430T	K	10½	9/10	2¾	5.00 × 15	9-24	114¾	55½	55½	3,000
420/430T	G	10½		2¾	5.00 × 15	9-24	114¾	55½	58¾	3,000
430T	LP	20		2¾	5.00 × 15	9-24	114¾	55½	58¾	
420/430RCU	K	10½	9/10	2¾	5.00 × 15	10-24	119¼	58¼	56	3,250
420/430RCU	G	10½		2¾	5.00 × 15	10-24	119¼	58¼	56	3,250
430RCU	LP	20		2¾	5.00 × 15	10-24	119¼	58¼	56	
420/430C	K	10½	9/10	2½	tracks—12″ shoes		102	56	51⅞	4,150 4R
420/430C	G	10½		2½	tracks—12″ shoes		102	56	51⅞	4,700 5R
430C	LP	20		2½	tracks—12″ shoes		102	56	51⅞	
430HC	K	10½	9/10	2¾	6.50 × 16	10-38	132	72	86*	3,400
430HC	G	10½		2¾	6.50 × 16	10-38	132	72	86*	3,400
430HC	LP	20		2¾	6.50 × 16	10-38	132	72	86*	
520/530	K	15½	2	4½	5.50 × 16	12.4-36	132¾	86⅝	83⅜*	4,960
520/530	G	18		4½	5.50 × 16	12.4-36	132¾	86⅝	83⅜*	4,960
520/530	LP	24		4½	5.50 × 16	12.4-36	132¾	86⅝	83⅜*	5,250
620/630RC	K	20½	1	6½	6.00 × 16	12.4-38	135¼	86⅝	88⅛*	5,858
620/630RC	G	22¼		6½	6.00 × 16	12.4-38	135¼	86⅝	88⅛*	5,858
620/630RC	LP	33		6½	6.00 × 16	12.4-38	135¼	86⅝	88⅛*	6,158
620/630HC	K	20½	1	6½	7.50 × 20	12.4 × 38	147 3/16	74	96¾*	7,690
620/630HC	G	22¼		6½	7.50 × 20	12.4 × 38	147 3/16	74	96¾*	7,690
620/630HC	LP	33		6½	7.50 × 20	12.4 × 38	147 3/16	74	96¾*	7,870
620/630S	K	20½	1	6½	6.00 × 16	13-30	126⅝	86⅝	81 7/16*	6,480
620/630S	G	22¼		6½	6.00 × 16	13-30	126⅝	86⅝	81 7/16*	6,480
620/630S	LP	33		6½	6.00 × 16	13-30	126⅝	86⅝	81 7/16*	6,780
620 grove	K	20	1	7	6.00 × 16	14-26	125⅞	77⅝	59½*	6,345
620 grove	G	20		7	6.00 × 16	14-26	125⅞	77⅝	59½*	6,345
620 grove	LP	38		7	6.00 × 16	14-26	125⅞	77⅝	59½*	6,645
720/730RC	K	24½	1	7⅛	6.00 × 16	12-38	135¼	86⅝	88¼*	6,790
720/730RC	G	26½		7⅛	6.00 × 16	12-38	135¼	86⅝	88¼*	6,790
720/730RC	LP	33		7⅛	6.00 × 16	12-38	135¼	86⅝	88¼*	7,100
720D-730D RC	D	20	¼	7**	6.00 × 16	12-38	135¼	86⅝	88¼*	7,390
720/730HC	K	24½	1	7⅛	7.50 × 20	13.6 × 38	148 3/16	74	101*	8,070
720/730HC	G	26½		7⅛	7.50 × 20	13.6 × 38	148 3/16	74	101*	8,070
720/730HC	LP	33		7⅛	7.50 × 20	13.6 × 38	148 3/16	74	101*	8,230
720D/730D HC	D	20	¼	7**	7.50 × 20	13.6 × 38	148 3/16	74	101*	8,470
720/730S	K	20½	1	7⅛	6.50 × 18	14-30	130¼	86⅝	87⅞*	7,380
720/730S	G	26½		7⅛	6.50 × 18	14-30	130¼	86⅝	87⅞*	7,380
720D/730S	LP	33		7⅛	6.50 × 18	14-30	130¼	86⅝	87⅛*	7,690
720D/730S	D	20	¼	7**	6.50 × 18	14-30	130¼	86⅝	87⅛*	7,790
820/830	D	32½	¼	8	7.50 × 18	15-34	142¾	79½	81*	7,850
435	D	10½		2½	5.00 × 15	10-34	136⅛	85¾	71.5*	3,750

**6¼ gallons with direct electric starting